DEFENSE, SECURITY AND STRATEGIES

EXPLOSIVE ORDNANCE DISPOSAL

ASSESSMENT AND ROLE OF EOD FORCES

DEFENSE, SECURITY AND STRATEGIES

Additional books in this series can be found on Nova's website under the Series tab.

Additional E-books in this series can be found on Nova's website under the E-book tab.

DEFENSE, SECURITY AND STRATEGIES

EXPLOSIVE ORDNANCE DISPOSAL

ASSESSMENT AND ROLE OF EOD FORCES

ALAIN MICHAUX
EDITOR

NOVINKA

New York

For permission to use material from this book please contact us:
Telephone 631-231-7269; Fax 631-231-8175
Web Site: http://www.novapublishers.com

Additional color graphics may be available in the e-book version of this book.

Library of Congress Cataloging-in-Publication Data

ISBN: 978-1-62808-357-6

Published by Nova Science Publishers, Inc. † New York

Contents

PREFACE

The Department of Defense (DOD)has relied heavily on the critical skills and capabilities of Explosive Ordnance Disposal (EOD) personnel from each of the four military services to counter threats from improvised explosive devices (IED), a significant cause of fatalities among U.S. troops in Iraq and Afghanistan. EOD personnel have extensive training in the detection, identification, on-site evaluation, making safe, recovery, and final disposal of unexploded explosive ordnance. EOD forces' capabilities in countering the IED threat, including collecting and evaluating captured explosive-related enemy material from the devices, have made these forces integral to successful joint military operations. However, the high demand for the EOD capability has resulted in personnel experiencing numerous deployments. In addition to their function in countering IEDs, EOD personnel are responsible for a wide range of other missions, such as clearing unexploded ordnance from training ranges; providing defense support to civil authorities; and assisting the U.S. Secret Service and Department of State with the protection of the President and other high-ranking government officials. This book provides an overview and assessment of the role of EOD forces and operations.

Chapter 1 - DOD has relied heavily on the critical skills and capabilities of EOD forces to counter the threat from improvised explosive devices on battlefields in Iraq and Afghanistan. The House Armed Services Committee directed DOD to submit a report on EOD force structure planning and directed GAO to review DOD's force structure plan. DOD's report provided little detail. GAO examined to what extent (1) DOD and the services have addressed increased demands for the EOD capability and identified funding to meet future requirements; and (2) DOD has developed guidance for employing the EOD capability effectively in joint operations. GAO evaluated DOD's report and EOD guidance; analyzed data on EOD missions, personnel, and funding; and interviewed DOD and service officials to gain perspectives from EOD personnel and managers.

Chapter 2 - The Army Explosive Ordnance Disposal program can trace its roots back to April 1941. The United States was not yet at war, but was actively preparing for that eventuality. Embassy personnel and military observers were reporting on the actions of warring nations and as these reports were evaluated by the War Department, Intelligence Sections, recommendations were made concerning actions that should be taken by the United States. Delayed-explosion bombs were creating havoc in Europe, taking a heavy toll on lives and industry. It was expected that if the United States entered the war, we would experience bombing of our cities and industries. As a result, the need for a bomb disposal program in this country received immediate attention. It was decided that both military and civilian bomb disposal personnel would be trained by the Army. All responsibility for bomb disposal was placed under the U.S. Army Ordnance Department.

Army Tactics, Techniques, and Procedures (ATTP) 4-32 provides the doctrinal guidance for commanders, staffs, and leaders at all levels responsible for explosive ordnance disposal (EOD) operations. The doctrine in this ATTP is based on suggestions, insights, and observations developed from past and current operations and an analysis of the future modular force support in line with national military strategy. It provides staff planning, responsibilities, capabilities, and requirements relevant to the EOD mission.

This manual focuses on the employment of EOD forces. This ATTP explains the mission and organization of EOD units. It describes why EOD forces are an important part of the combined arms team throughout all phases of full spectrum operations. It also prescribes responsibilities for EOD forces in support of the Army, Joint Force, and subordinate commanders. The manual reflects and supports the Army operations doctrine as stated in Army Doctrine Publication (ADP 3-0).

This ATTP provides commanders and staff with doctrine for EOD operations and integration of EOD forces. It describes the EOD functional relationships which integrate EOD with the warfighting functions. This ATTP serves to illustrate Army EOD force structure operating as an enabler in the joint operating environment.

This publication applies to the Active Army, the Army National Guard (ARNG)/Army National Guard of the United States (ARNGUS), and the United States Army Reserve (USAR) unless otherwise stated.

The proponent of this publication is the United States Army Training and Doctrine Command (TRADOC). Send comments and recommendations on DA Form 2028 (Recommended Changes to Publications and Blank Forms [or in DA Form 2028 format]) directly to Commander, Combined Arms Support Command, Directorate of Training and Doctrine, ATTN: ATCL-O, 2221 Adams Avenue, Fort Lee, VA 23801-1809.

In: Explosive Ordnance Disposal
Editor: Alain Michaux
ISBN 978-1-62808-357-6

Chapter 1

EXPLOSIVE ORDNANCE DISPOSAL: DOD NEEDS BETTER RESOURCE PLANNING AND JOINT GUIDANCE TO MANAGE THE CAPABILITY*

United States Government Accountability Office

ABBREVIATIONS

DOD	Department of Defense
EOD	Explosive Ordnance Disposal
IED	Improvised Explosive Devices

WHY GAO DID THIS STUDY

DOD has relied heavily on the critical skills and capabilities of EOD forces to counter the threat from improvised explosive devices on battlefields in Iraq and Afghanistan. The House Armed Services Committee directed DOD to submit a report on EOD force structure planning and directed GAO to

* This is an edited, reformatted and augmented version of The United States Government Accountability Office publication, Report to Congressional Committees GAO-13-385, dated April 2013.

review DOD's force structure plan. DOD's report provided little detail. GAO examined to what extent (1) DOD and the services have addressed increased demands for the EOD capability and identified funding to meet future requirements; and (2) DOD has developed guidance for employing the EOD capability effectively in joint operations. GAO evaluated DOD's report and EOD guidance; analyzed data on EOD missions, personnel, and funding; and interviewed DOD and service officials to gain perspectives from EOD personnel and managers.

What GAO Recommends

To better enable DOD to plan for funding EOD mission requirements and enhance future use of EOD forces in joint combat operations, GAO recommends that DOD direct (1) the Secretaries of the Army and the Navy to collect data on current Army and Marine Corps EOD funding, and (2) the Chairman of the Joint Chiefs of Staff to develop joint EOD doctrine that would guide combatant commanders' planning and clarify joint operational roles and responsibilities. In oral comments on a draft of this report, DOD concurred with the recommendations.

What GAO Found

Explosive Ordnance Disposal (EOD) forces grew over the past 10 years to meet wartime and other needs, but the Department of Defense (DOD) does not have the data needed to develop a funding strategy to support future EOD force plans. To meet increased demands for EOD personnel, the services increased their EOD forces from about 3,600 personnel in 2002 to about 6,200 in 2012. Anticipating that the need for EOD will continue as forces withdraw from ongoing operations, the services intend to maintain their larger size. The Navy and Air Force have data on the baseline costs for some or all of their EOD activities, but the Army and Marine Corps do not have complete data on spending for EOD activities. Therefore, DOD does not have complete data on service spending on EOD activities needed to determine the costs of its current EOD capability and to provide a basis for future joint planning. Until all the services have complete information on spending, service and DOD leadership will be unable to effectively identify resource needs, weigh priorities, and assess budget trade-offs.

EOD forces from all four services have worked together in Iraq and Afghanistan and the services have developed guidance on tactics and procedures for EOD forces, but challenges persist because DOD has not institutionalized joint EOD doctrine through a joint publication. Joint doctrine facilitates planning for operations and establishes a link between what must be accomplished and the capabilities for doing so. DOD studies have noted commanders' limited awareness of EOD capabilities during combat operations, and EOD personnel reported challenges they attributed to non-EOD forces' lack of understanding of EOD operations. Several DOD organizations have responsibilities for some EOD functions, but no entity has been designated as the focal point for joint EOD doctrine. Joint doctrine could help leaders identify EOD capability requirements and better position combatant commanders in their use of EOD forces in future operations. Joint doctrine that is developed and approved as authoritative guidance would enhance the EOD forces' ability to operate in an effective manner, and would better position the services to identify capability gaps in meeting service, joint, and interagency requirements; to invest in priority needs; and to mitigate risks.

April 25, 2013

Congressional Committees

The Department of Defense (DOD) has relied heavily on the critical skills and capabilities of Explosive Ordnance Disposal (EOD) personnel from each of the four military services to counter threats from improvised explosive devices (IED), a significant cause of fatalities among U.S. troops in Iraq and Afghanistan.[1] EOD personnel have extensive training in the detection, identification, on-site evaluation, making safe, recovery, and final disposal of unexploded explosive ordnance. EOD forces' capabilities in countering the IED threat—including collecting and evaluating captured explosive-related enemy materiel from the devices— have made these forces integral to successful joint military operations. However, the high demand for the EOD capability has resulted in personnel experiencing numerous deployments. In addition to their function in countering IEDs, EOD personnel are responsible for a wide range of other missions, such as clearing unexploded ordnance from training ranges; providing defense support to civil authorities; and assisting the U.S. Secret Service and Department of State with the protection of the President and other high-ranking government officials.

In House Report 112-78, which accompanied a House bill for the National Defense Authorization Act for Fiscal Year 2012, the House Armed Services Committee directed the Secretary of Defense to submit a report on Explosive Ordnance Disposal force structure planning to the congressional defense committees by March 1, 2012. [2] Further, the committee directed GAO to review DOD's force structure plan and to report our findings to the congressional defense committees.[3] We reviewed and evaluated the DOD report on EOD force structure directed by the committee, and in July 2012 we briefed the House Armed Services Committee staff on our observations of DOD's force structure plan. We found that DOD's classified one-page EOD report provided limited detail or context about the status of the department's EOD force or plans for future EOD capability requirements. Also, we found that DOD had not established a new consolidated budget justification display that fully identified the services' baseline EOD budgets, as directed by the House Armed Services Committee.[4]

In order to provide additional information on the status of DOD's EOD forces and identify baseline EOD budgets, we gathered and analyzed data on EOD forces' support for joint operations and the services' plans to support future EOD requirements. We reviewed and analyzed prior DOD reports on the EOD force, collected available data on personnel[5] and on budgets, and interviewed cognizant officials from the Office of the Secretary of Defense and the services. In addition, we conducted 28 group discussion sessions with EOD personnel to obtain their perspectives on the EOD capability. Specifically, we examined the extent to which (1) DOD and the services have addressed increased demands for the EOD capability and identified funding to meet anticipated future requirements; and (2) DOD has developed guidance for employing the EOD capability effectively in joint operations.

To determine the extent to which DOD and the services have addressed increased demands for the EOD capability, we collected and analyzed data from each of the four services' EOD forces, including data on the organizational structure of EOD forces; on the size of EOD forces since 2002; and on projected manpower needs. Further, we interviewed key DOD EOD officials across the department—including from the Office of the Secretary of Defense, each of the services, and various support commands—to gain their perspectives on the operational tempo of EOD forces; the use of joint EOD forces for recent EOD combat missions, and the challenges EOD forces experienced; and expected future EOD mission requirements. To determine the extent to which DOD and the services have identified funding to resource EOD forces to meet anticipated future requirements, we collected and analyzed available EOD funding data from each of the services for the past 3 fiscal years.

To determine the extent to which DOD has developed guidance for employing the EOD capability effectively in joint operations, we reviewed existing service, multi-service, and joint DOD regulations and doctrine; reviewed and analyzed prior DOD reports' findings about doctrine; and determined whether associated recommendations had been implemented. We interviewed key DOD EOD officials to ascertain the extent to which DOD has comprehensive joint EOD guidance. Additionally, we met with officials from selected EOD units in all four military services and conducted group discussions with EOD-qualified team members, team leaders, senior enlisted personnel, and officers to obtain their perspectives on issues related to military service-specific missions, joint combat operations, training, equipment, and operational tempo. A more detailed description of our scope and methodology can be found in appendix I.

We conducted this performance audit from May 2012 through April 2013 in accordance with generally accepted government auditing standards. Those standards require that we plan and perform the audit to obtain sufficient, appropriate evidence to provide a reasonable basis for our findings and conclusions based on our audit objectives. We believe that the evidence obtained provides a reasonable basis for our findings and conclusions based on our audit objectives.

BACKGROUND

As depicted in figure 1 below, the services maintain highly trained EOD personnel responsible for eliminating explosive hazards in support of a range of events, from major combat operations and contingency operations overseas to range clearance to protecting designated persons, such as the President of the United States. The services' EOD forces are dispersed worldwide to meet combatant commanders' requirements. Units may be deployed together or organized into smaller teams, as missions require. EOD technicians generally work in two- or three-person teams to identify and disarm ordnance.

Servicemembers volunteer[6] for the EOD force and attend joint basic EOD training at the Naval School Explosive Ordnance Disposal at Eglin Air Force Base in Florida. This 143-day course—staffed by Army, Navy, Air Force, and

Marine Corps instructors—provides instruction in munitions identification, render-safe procedures, explosives safety, and EOD-unique equipment. Servicemembers who successfully complete training are certified as Basic EOD Technicians and, afterward, join their EOD units. This process is shown in figure 2 below. With additional training and experience, a Basic EOD Technician can earn advanced certifications as a Senior EOD Technician and a Master EOD Technician.

Source: Defense Video and Imagery Distribution System.

Figure 1. EOD Personnel Supporting a Range of Events.

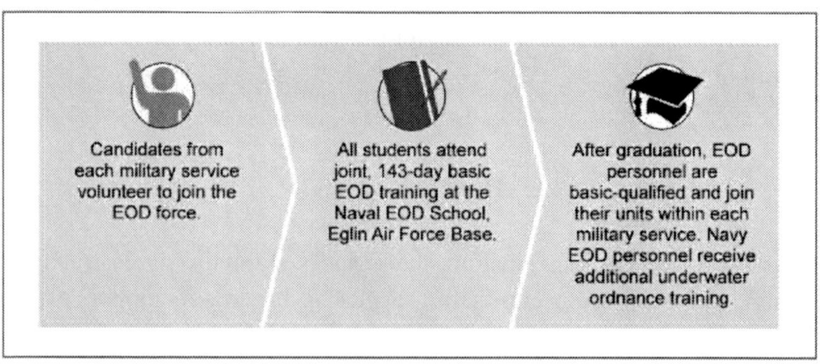

Source; GAO analysis of DOD information.

Figure 2. Process for Becoming Certified as a Basic EOD Technician.

The EOD capability is generally viewed as an activity that–comparable with logistics—supports combat forces, rather than as a combat activity. EOD forces in the Army, Air Force, and Marine Corps are considered support activities and are organizationally placed with other support forces. Examples of where EOD forces are organizationally aligned include the ordnance corps (Army), the engineer forces (Air Force and Marine Corps), and aviation forces (Marine Corps). Previous DOD studies have noted that this organizational alignment of EOD forces created operational challenges in recent operations. For example, one study noted that EOD forces, as support forces, were not perceived as available in some phases of a battle. In contrast with the other services, the Navy identifies its EOD force as a combat activity and it is organizationally placed within the Navy Expeditionary Combat Command.[7] According to Navy officials, this better reflects the battlefield role of its EOD force and provides advocacy at senior levels in the Navy.

Other organizations have responsibilities for EOD activities and policy. Within the Office of the Secretary of Defense, the Assistant Secretary of Defense for Special Operations and Low Intensity Conflict serves as the Office of the Secretary of Defense proponent for EOD and is responsible for developing, coordinating, and overseeing the implementation of DOD policy for EOD technology and training. The Joint Staff's Force Protection Division (J8 Directorate) oversees requirements from combatant commands related to the department's EOD capability. The Secretary of the Navy is the DOD single manager for joint service EOD technology and training.[8] The Navy is supported in this role by a joint EOD Program Board made up of general or flag officers from each of the services who function as their respective services' focal points for EOD program requirements. The Program Board establishes the joint EOD program and approves the plan and budget. The Technical Training Acceptance Board, under the Program Board, coordinates, approves, and standardizes all EOD common-type individual training.[9] The Program Board also has a role in prioritizing and recommending funding for research and development of equipment based on joint EOD requirements.

EOD Forces Grew to Meet Wartime and Other Demands, But DOD Does Not Have Data Needed to Develop Funding Strategy to Support Future EOD Force Plans

To meet increased demands for EOD personnel, the services increased the size of their EOD forces. Based on available data, we determined that the services grew from about 3,600 personnel in 2002 to about 6,200 in 2012—an almost 72 percent increase. The services anticipate that even as forces withdraw from recent operations the need for EOD forces will continue, so they intend to maintain their larger size. However, the respective services' abilities to identify and track spending on EOD activities vary, so DOD does not have complete information on EOD spending. The House Armed Services Committee directed DOD to establish a consolidated budget justification display fully identifying the services' baseline EOD budgets, but DOD has not done so. Without complete EOD spending information, the services and DOD may have difficulty in justifying the future EOD force structure, and in informing future funding plans and trade-offs among competing priorities.

Services Increased Numbers of EOD Personnel to Meet Demand for the Capability

Over the past decade of military operations, the services all took actions to increase their EOD capabilities. As figure 3 shows, the Army more than doubled its EOD forces, with the largest increases occurring after fiscal year 2006. The Marine Corps and the Navy increased their EOD forces by approximately 77 percent and 20 percent, respectively. The Air Force increased its EOD forces by approximately 36 percent.[10] The services anticipate maintaining EOD personnel numbers at the 2012 level at least through the next 5 years, as is also shown in figure 3, although final decisions on EOD force size and structure will depend on future DOD budgets. According to DOD EOD officials, the time required to train qualified EOD personnel is lengthy; therefore, EOD is not a capability that can be built up quickly.

Meeting the demands for EOD forces in combat operations has negatively affected EOD units' personnel and ability to train for other missions. Each of the services met the combatant commanders' high demand for EOD personnel by deploying EOD units and assuming risks as other EOD missions and training activities were left unfulfilled. For example, according to service officials, the services assumed some risks in some mission areas such as countering sea mines, clearing unexploded ordnance on training ranges, and providing defense support to civil authorities. EOD personnel who participated in our group discussions said they experienced multiple deployments and limited time at home, maintaining a pace that was exacerbated by time spent away from home in training and in support of the U.S. Secret Service and Department of State in the protection of important officials.[11] (Appendix II summarizes, in greater detail, the operational and personnel issues raised by EOD personnel who participated in group discussions we held.)

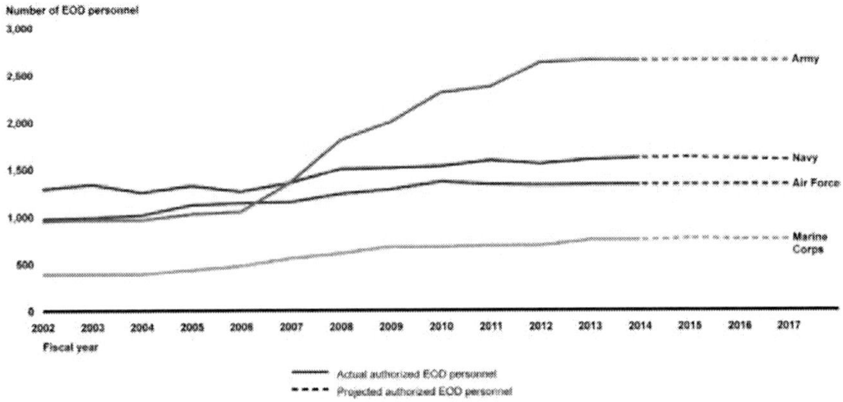

Source: GAO analysis of service manning data.

Figure Notes: Authorized personnel refers to the manning level the military services authorized for their EOD forces, as presented in service manning data. The Army and Air Force data above do not include EOD officers authorized prior to 2005 and 2008, respectively, because service officials could not provide those numbers.

Figure 3. EOD Personnel Growth by Military Service, Fiscal Years 2002 through 2017.

As the services begin turning their focus away from training for deployments to Afghanistan to counter IEDs, officials from across the services and DOD officials noted that the services will need to retain the current EOD size so that they can expand training for their core missions and prepare for future requirements. For example, according to Navy officials, Navy EOD personnel will re-emphasize skills—such as diving— that are needed for their core missions. EOD forces will also be assigned to traditional missions to fill gaps in capabilities that came about when EOD forces were deployed to Afghanistan and Iraq. For example, according to DOD officials, EOD personnel will be available to combatant commanders for humanitarian demining, irregular warfare, and building international partner capacity activities. Also, according to Army officials, Army EOD forces will be available to respond more quickly to incidents involving unexploded military ordnance found in local communities. In addition, EOD forces will continue to provide support to the Very Important Persons Protection Support Activity's mission of ensuring the safety of federal officials, such as the President, as they travel. In fiscal year 2012, the services provided more than 473,000 hours of support to this activity; as a presidential election year, 2012 required greater than the annual average of about 300,000 hours provided from fiscal years 2007 through 2011.

DOD officials expect EOD capabilities to continue to be in demand by combatant commanders for the foreseeable future. For example, officials believe that the IED threat is likely to persist given its low cost and high accessibility to non-state adversaries. In addition, based on the primary missions highlighted in DOD's current strategic guidance,[12] the services anticipate continued requirements for EOD capabilities. For example, EOD capabilities are expected to be needed for several missions including, among others: (1) countering terrorism and irregular warfare; (2) countering anti-access/area denial measures, including mining; (3) countering weapons of mass destruction; and (4) providing support to civil authorities. The mission of countering anti-access/area denial measures, in particular, will require Navy EOD forces to train to counter anti-access measures that use sea mines and to clear explosive obstacles in sea lines of communication. Also, Air Force EOD forces will be expected to train to support recovery operations to keep air bases and runways clear of unexploded ordnance. All the services' EOD forces will need to continue to train for homeland missions of providing support to civil authorities. In addition, EOD forces' ability to conduct humanitarian demining activities can support combatant commanders' efforts to help build relationships with other countries, according to DOD officials.

Lack of Full Visibility into EOD Funding Complicates Planning for Resources to Meet Future Requirements

DOD does not have the consolidated information on service spending on EOD activities needed to enable it to determine the amount of resources currently being devoted to EOD capabilities and how the services are planning to support joint EOD capability needs. The House Armed Services Committee directed DOD to establish a consolidated budget justification display that fully identifies the military services' baseline EOD budgets and encompasses all programs and activities of the EOD force for specific functions.[13] Further, GAO's *Standards for Internal Control in the Federal Government*[14] state that managers need financial data to determine whether they are meeting their accountability goals for effective and efficient use of resources. However, DOD has not taken steps to identify comprehensive financial data on EOD activities across the services, nor has it responded to direction by the House

Armed Services Committee to establish a new consolidated budget justification display that fully identifies the services' baseline EOD budgets and encompasses all programs and activities of the EOD force for the functions of procurement; operation and maintenance; and research, development, testing, and evaluation. According to an Office of the Under Secretary of Defense (Comptroller) official, the DOD budget is not constructed or structured to report at the level of detail called for by the committee.

In the absence of a departmental budget justification for EOD activities, we sought to compile DOD's EOD spending but we were unable to collect comprehensive and reliable data because the completeness of the services' EOD budget data varied. We requested funding data from all the services for EOD activities for fiscal years 2010 through 2012. According to the data the services provided, they received funding for EOD activities from their regular funding accounts for military personnel, for procurement, and for operation and maintenance. In addition, Congress has provided overseas contingency operations appropriations to the services for EOD activities. Further, the Joint IED Defeat Organization— which is largely funded by overseas contingency operations appropriations—paid for or provided some equipment, such as robots, and training, such as courses on the design and construction of IEDs. [15] However, we found that the Army and Marine Corps had incomplete or little available information but information from the Navy and Air Force was more complete. Specifically, we found the following:

- The Army, which has the largest EOD force, does not have full visibility over EOD funding. Army officials collected some information on funding at our request, but they could not be sure that the data were complete and accurate. The Army regulation on EOD assigns responsibility for monitoring funding for the Army EOD program to the Deputy Chief of Staff for Operations, Plans, and Training (G-3/5/7).[16] Officials in that office, however, told us they do not have access to complete funding information because funding for EOD activities is spread across multiple programs, functions, or organizations. Officials who oversee the EOD program told us they would like to have funding data to assist in managing and prioritizing the Army's EOD operations, but they have no plans currently to collect it.
- The Marine Corps EOD program officials could readily provide us with procurement funding information, but not with comprehensive

information on funding that could have come from other funding accounts, such as military personnel and operation and maintenance.

- Officials in the office of the Navy EOD program resource sponsor could provide funding information because the Navy has a dedicated EOD program element code17 and it tracks funding for the EOD capability separately on a continuous basis to enable it to manage its own capability.
- Officials in the Air Force EOD program oversight organization could provide funding information on operating, maintaining, and procuring equipment and other items for the Air Force EOD force that it compiles and uses to manage the Air Force EOD program. However, it could not readily provide information on military personnel because, according to officials, personnel are accounted for across more than 30 program elements. Starting in fiscal year 2013, the Air Force began to use a dedicated EOD program element code to enable better identification of current EOD spending and to provide justification for the EOD capability in future budget requests.

The services anticipate maintaining their currently authorized EOD personnel levels.[18] However, planning for the future EOD capability may be hampered by DOD's lack of visibility into the current costs of EOD capabilities across the services. According to officials from each service, overseas contingency operations funding has been used to provide equipment and training to EOD forces for the past several years, such as that shown in figure 4 below. In the future these costs will have to be funded from regular appropriations, and to compete with other service priorities.19 For example, the services received EOD robots and mine resistant ambush-protected vehicles from overseas contingency operations funding. Should EOD units need to continue to use the equipment in the future or to acquire similar such equipment, maintaining and procuring it may have to be funded through regular appropriations. Overseas contingency operations funding also provided advanced homemade explosive, forensic, and medical training opportunities that EOD technicians in our discussion groups thought were valuable to their missions and their safety.

Service officials expressed concerns to us about the adequacy of future funding for their EOD forces after overseas contingency operations funding is phased out, but the extent to which the services have identifiable funding plans for future EOD activities varied.

Source: Defense Video and Imagery Distribution System.

Figure 4. Example of Homemade Explosive Training Course and Mine Resistant Ambush-Protected Vehicle.

The Navy and Air Force now have program element codes that enable service officials to identify and evaluate the appropriate level of spending on their EOD capabilities. However, the Army's and Marine Corps' lack of complete data on the costs of their current EOD forces negatively affects their efforts to develop viable funding plans for supporting their EOD capability into the future. Until the services have information on current spending as well as justification for their future funding needs, service and DOD leadership will be unable to effectively identify resource needs, weigh priorities, and assess budget trade-offs within anticipated declining resources. Moreover, the lack of visibility into current spending and future funding plans may impede DOD's ability to provide Congress with information needed to facilitate its oversight.

EOD Forces Have Worked Together in Joint Operations, but DOD Has Not Developed Joint EOD Doctrine to Guide Effective Employment of the Capability

EOD forces have operated jointly in Iraq and Afghanistan to fulfill battlefield requirements, and the services have jointly developed guidance on tactics, techniques, and procedures for EOD forces,[20] but DOD has not fully institutionalized the guidance through joint EOD doctrine in the form of a Joint Publication. According to DOD, the purpose of joint doctrine is to enhance the operational effectiveness of U.S. joint forces.[21] It is written for those such as the services, among other recipients, who prepare and train forces for carrying out joint operations. Joint doctrine facilitates planning for and execution of operations, and it establishes a link between what must be accomplished and the capabilities for doing so by providing information on how joint forces can achieve military strategic and operational objectives in support of national strategic objectives. According to service EOD officials, joint doctrine also provides standardized terminology. EOD personnel and officials told us, however, that they had encountered repeated challenges—such as a lack of planning for EOD capabilities as well as variations among the services' procedures—during joint combat operations. A key reason for the services' challenges is the absence of a consistent understanding of EOD operations, including expectations for how forces should plan operations and work together. The services are disadvantaged with respect to EOD capabilities, knowledge, and use because DOD has not developed joint doctrine in the form of a Joint Publication.

The Services Use Common EOD Operational Guidance, but Challenges Persist in Operating Jointly

In 2001, DOD's Air Land Sea Application Center [22] published multi-service guidance outlining a set of EOD tactics, techniques, and procedures for employing EOD forces jointly in a range of military operations.[23] The guidance, which was updated in 2005 and 2011, applies to leaders, planners, and EOD personnel and provides information that can help them understand each military service's capabilities. The multi-service guidance, as updated, has been in place for more than a decade, but challenges in joint EOD combat operations have continued, as prior DOD studies on EOD capabilities have reported. One study described EOD forces' efforts to work jointly as being "somewhat ad hoc" and noted that culture, technique, and language differences among the military services caused challenges in working together.[24] Another study reported on some combat unit commanders' limited awareness of EOD capabilities during overseas combat operations.[25] For example, in some instances Army combat unit commanders did not understand the differences between the capabilities of EOD personnel who are trained to safely disarm ordnance and the capabilities of Army combat engineers, some of whom have limited training on disposing of select unexploded ordnance. As a result, some combat engineers, who are not trained to safely disarm ordnance, destroyed ordnance caches, resulting in improper handling of some of the ordnance caches and creating a more dangerous site containing debris as well as potentially unexploded ordnance.

EOD personnel we spoke with reported experiencing similar challenges during recent deployments in support of joint combat operations, and as the observations recorded below indicate, they attributed these challenges to a lack of understanding of EOD operations on the part of non-EOD forces. EOD personnel indicated the following:

- *Commanders of combat units did not always take into account differences among the military services' EOD forces.* For example, Marine Corps EOD personnel commented that their EOD standard operating procedures include conducting dismounted patrols, while at one time the Army's EOD personnel were not allowed to dismount from vehicles to conduct EOD operations. These Marine Corps personnel stated that Marine Corps commanders were unsure how to work with Army EOD forces supporting Marine units. Similarly, Air Force personnel said they were not trained to dismount to search for explosives, as Marine Corps commanders expected them to be.
- *Non-EOD personnel in combat units did not always understand EOD protocols or capabilities.* For example, Army EOD personnel cited an instance in which a non-EOD officer at a forward operating base picked up post-blast fragments of an improvised explosive device at a blast site, thus disturbing the site and contaminating potential forensic evidence. Other examples include commanders not securing sites where unexploded improvised explosive devices were found, and non-EOD personnel tampering with unexploded devices in an attempt to deactivate them before EOD personnel arrived.
- *Requests for EOD support did not always take into account differences in how the various services' EOD forces are organized.* For example, an Army EOD Company contains approximately 40 people, while a Navy EOD Mobile Platoon has 8 people. According to Navy EOD personnel, battlefield commanders they supported sometimes received a smaller EOD force than was needed and expected, or conversely, sometimes received a larger force, which would require more logistical support than planned.

We found that pertinent Navy and Air Force regulations only briefly mention joint EOD operations, and they refer to the multi-service guidance as a source for more detailed information. However, the Army's and Marine

Corps' regulations do not discuss joint EOD operations in detail or refer to the multi-service guidance. A DOD study assessing the joint EOD capability reported that this multi-service guidance was not widely used by combatant commands.[26] In addition, according to DOD officials, this multi-service guidance is not as authoritative to the services and combatant commands as a Joint Publication, and is generally not used by the services to develop force requirements.

DOD Does Not Have Joint Doctrine in the Form of a Joint Publication to Provide Authoritative Doctrinal Guidance for Joint EOD Operations

Prior DOD studies have highlighted the need for joint doctrine on EOD operations and noted that current guidance is insufficient, but the Joint Staff has not published joint doctrine for EOD operations. We found that several DOD doctrinal joint publications refer to EOD activities, but most references are limited, recognizing the need for EOD but not providing additional guidance as to how such capabilities should apply to operations. EOD is briefly mentioned or discussed in greater detail in 30 unclassified or for official use only doctrinal joint publications. For example, the role of EOD is briefly mentioned in joint doctrine about operations such as antiterrorism, foreign humanitarian assistance, and evacuating noncombatants. The EOD role is more fully discussed in joint doctrine about countering the improvised explosive device threat, joint engineer operations, and addressing obstacles—such as unexploded ordnance and sea mines—that could be encountered by joint forces in a range of military operations. According to a Chairman of the Joint Chiefs of Staff manual on joint doctrine development, part of the development philosophy for joint doctrine is that it continues to evolve as the United States Armed Forces adapt to meet national security challenges.[27] However, none of these joint publications addresses the full range of EOD capabilities and potential activities. As previous assessments of DOD's joint EOD capabilities have reported, the lack of joint EOD-specific doctrine limits EOD forces' and planners' ability to identify and mitigate capability gaps.

According to the Chairman of the Joint Chiefs of Staff manual, joint doctrine projects must be formally sponsored by a service chief, a combatant commander, or a director of a Joint Staff directorate.[28] A key reason why joint EOD doctrine has not been developed is that no entity has been made accountable for following through on recommendations and sponsoring EOD doctrine, to include stakeholder coordination, for development of joint guidance. Although several organizations have responsibilities for some EOD functions, no one entity has been designated as the focal point on joint doctrine or operational issues. For example, the Assistant Secretary of Defense for Special Operations and Low Intensity Conflict, as the Office of the Secretary of Defense proponent for EOD, is charged with developing, coordinating, and overseeing the implementation of DOD policy for EOD technology and training, but that official is not involved with oversight of joint doctrine or operational issues. Similarly, the joint EOD Program Board, chaired by a flag officer designated by the Secretary of the Navy and comprising general officer representation from the other services, is generally focused on joint common EOD technology and training issues. One DOD EOD study, which according to a Joint Staff official was initiated by a former Vice Chairman of the Joint Chiefs of Staff, recommended among other things that the Joint Staff sponsor the development of joint EOD doctrine. However, during DOD's review process that recommendation was sent back to the Joint Staff for reconsideration, where, according to a Joint Staff official, the matter was dropped and never presented to senior leaders.

Having joint guidance, such as joint doctrine, could put combatant commanders in a better position to make decisions about using EOD forces in future operations. In addition, joint EOD doctrine could provide a basis for planning and identifying capability requirements for future operations. Having joint doctrine that is developed and approved by the Chairman of the Joint Chiefs of Staff as authoritative guidance would enhance DOD's EOD forces' ability to operate in an effective manner and better position the military services to identify capability gaps in meeting service, joint, and interagency requirements; to invest in priority needs; and to mitigate risks.

Conclusion

As IEDs became a significant threat to U.S. forces in Iraq and Afghanistan, EOD emerged as a critical capability, and over the past decade, the services increased the size of their EOD forces. Growing the EOD capability takes time because of the highly technical training required and the additional experience needed to become proficient in handling dangerous unexploded ordnance. Looking toward the future, DOD and the services believe that broad demand for this capability will continue. Growth of the EOD forces until now has been funded in part by overseas contingency operations dollars, but these funds are likely to decrease as operations in Afghanistan diminish. A major challenge facing the EOD community, especially the Army and Marine Corps, is the lack of complete information to clearly show the resources it will take to sustain their larger force levels. Further, DOD does not have good visibility into service spending on EOD forces. Without comprehensive information on the costs of the services' EOD forces, senior service and DOD leaders are not well positioned to justify the current EOD force structure or to ensure that funding goes to priorities in accordance with strategic guidance. In addition, the absence of comprehensive information limits DOD's ability to respond to congressional requests for budget information and may continue to hamper Congress' oversight of the health and viability of the EOD force.

Although EOD forces from each of the services have deployed together to support recent ground operations, attention to EOD as a joint capability has been limited. Differences among the services' procedures complicate joint force planning and operations, and there is little common understanding of the EOD capability outside of the EOD force. The lack of understanding of EOD capabilities among those battlefield commanders has caused them challenges in providing the right capabilities to maximize the effectiveness of operations to protect U.S. forces and to collect information to defeat the networks of insurgents using IEDs against U.S. forces. A number of publications refer to EOD and its capabilities for specific functions, but none provides clear and complete guidance for integrating the activities of EOD forces with other combat activities and maximizing the capabilities that EOD forces provide. In addition, no entity has followed through on previous recommendations to sponsor and advocate for developing joint EOD doctrine. With joint doctrine that specifies the role of EOD in joint operations and provides a consistent lexicon for joint planning, EOD participation in joint operations could be more efficient and effective. In addition, with joint doctrine regarding future requirements for the joint EOD capability, the services will have more complete information to inform their force structure planning and provide adequately trained and experienced forces to meet future requirements.

RECOMMENDATIONS FOR EXECUTIVE ACTION

We recommend that the Secretary of Defense take the following two actions:

- To improve the Army's and Marine Corps' ability to ensure adequate support of their EOD forces within expected budgets, direct the Secretaries of the Army and the Navy to collect data on costs associated with supporting their current EOD forces.
- To enhance the future employment of EOD forces in joint combat operations, direct the Chairman of the Joint Chiefs of Staff to develop joint EOD doctrine that would guide combatant commanders' planning and clarify joint operational roles and responsibilities.

Agency Comments and Our Evaluation

We provided a draft of this report to the Secretary of Defense for comment. An official from the Office of the Assistant Secretary of Defense for Special Operations and Low Intensity Conflict provided oral comments on the draft indicating that DOD concurred with our report and both of our recommendations.

John H. Pendleton Director
Defense Capabilities and Management

Appendix I: Scope and Methodology

The scope of our review on Explosive Ordnance Disposal (EOD) forces included the Office of the Secretary of Defense, the Joint Staff, and the military services, including select EOD units from each service, and other DOD organizations that utilized or impacted EOD forces.

We obtained relevant documentation and interviewed key officials from the following offices:

- Office of the Assistant Secretary of Defense for Special Operations and Low-Intensity Conflict;
- The Joint Staff
  - J34 – Deputy Directorate for Antiterrorism/Homeland Defense;
  - J5 – Strategic Plans and Policy Directorate; and
  - J8 – Force Structure, Resources, and Assessment Directorate;
- Department of the Army

- Office of the Deputy Chief of Staff (Operations, Plans, and Training) EOD & Render Safe Procedures Branch;
- U.S. Army Ordnance Corps Explosive Ordnance Disposal Directorate, Fort Lee, Virginia;
- 20th Support Command, Aberdeen Proving Ground, Maryland;
- 52nd Ordnance Group (EOD), Fort Campbell, Kentucky;
    - 49th EOD Company, Fort Campbell, Kentucky;
    - 723rd EOD Company, Fort Campbell, Kentucky; and
    - 788th EOD Company, Fort Campbell, Kentucky;
- Department of the Navy

Navy

- Office of the Chief of Naval Operations, Navy Expeditionary Combat Requirements Branch;
- Navy Expeditionary Combat Command, Joint Expeditionary Base Little Creek-Fort Story, Virginia;
    - EOD Group One, Naval Amphibious Base Coronado, California;
        - EOD Mobile Unit One, Naval Base Point Loma, California,
        - EOD Mobile Unit Three, Naval Amphibious Base Coronado, California,
        - EOD Mobile Unit Eleven, Imperial Beach, California,
        - EOD Training and Evaluation Unit One, Naval Base Point Loma, California,
        - EOD Expeditionary Support Unit One, Naval Amphibious Base Coronado, California, and Mobile Diving and Salvage Unit One, Naval Amphibious Base Coronado, California;
- Executive Manager of EOD Technology and Training, Washington, D.C.
    - Naval EOD Technology Division, Naval Support Facility Indian Head, Maryland;
    - Naval School EOD, Eglin Air Force Base, Florida;
- Center for EOD and Diving, Naval Support Activity Panama City, Florida;

Marine Corps

- EOD Occupational Field Sponsor, Headquarters, U.S. Marine Corps;

    - 1st EOD Company, Camp Pendleton, California;
    - 3rd EOD Company, Okinawa, Japan; and
    - EOD Personnel Supporting the 3rd Marine Aircraft Wing, Marine Corps Air Station Miramar, California;
- Department of the Air Force
    - EOD Program Directorate;
    - Air Force Civil Engineer Support Agency, Tyndall Air Force Base, Florida;
    - 96th Civil Engineering Squadron EOD Flight, Eglin Air Force Base, Florida;
- Joint Improvised Explosive Device (IED) Defeat Organization, Arlington, Virginia; and
- U.S. Northern Command Joint Force Headquarters, National Capitol Region, Joint EOD – Very Important Persons Protection Support Activity, Fort McNair, Washington, D.C.

To determine the extent to which DOD and the services addressed increased demands for the EOD capability, we collected and analyzed data and descriptions from each of the four services' EOD forces on their traditional military service EOD missions, including the total number of hours dedicated to support the Very Important Persons Protection Support Activity. After receiving the data showing the total number of hours dedicated to support this activity, we reviewed the data and interviewed the U.S. Northern Command official who provided them to assess the data's reliability. Based on these actions, we determined that these data were sufficiently reliable to include the amount of hours of support EOD personnel provided to this activity. We also requested data on the organizational structure of EOD forces from each of the services; operational tempo of EOD units and the authorized numbers of EOD officers and enlisted personnel within each service from fiscal year 2002 through fiscal year 2012; as well as projected manpower needs through fiscal year 2017. After receiving the authorized EOD personnel data, as identified by the services, we interviewed service officials who had provided it and other subject matter experts to assess the reliability of the data. Based on our review of the personnel data provided and our interviews, we determined that the personnel data were sufficiently reliable to describe growth in numbers of EOD personnel. Additionally, we interviewed key cognizant DOD officials with responsibility for EOD activities across the department—including from the Office of the Secretary of Defense, each of the military services, selected support commands, and units—to gain their perspectives on the operational tempo of EOD forces, use of joint EOD forces for recent EOD combat missions and challenges, and expected future EOD mission requirements.

To determine the extent to which DOD and the services have identified funding to resource EOD forces to meet anticipated future requirements, we collected and analyzed available EOD funding data from each of the services for fiscal years 2010 through 2012. We requested that the services provide EOD funding data from the base and overseas contingency operations budgets for specific funding accounts, including military personnel; operation and maintenance; procurement; and research, development, test, and evaluation. Also, we requested that the services identify funding, if any, received from the Joint IED Defeat Organization, or other sources. The comprehensiveness of the data provided by each of the military services varied. After receiving the funding data, we interviewed service officials who had provided them to assess the reliability of the data. Based on our review of the funding data provided and our interviews, we determined that the funding data were incomplete, potentially inaccurate, and not sufficiently reliable to establish a baseline level of DOD's EOD spending.

To determine the extent to which DOD has developed guidance for employing the EOD capability effectively in joint operations, we systematically analyzed 75 unclassified and for official use only documents of existing joint DOD doctrine (Joint Publications) to identify the inclusion of EOD functions. One GAO analyst conducted this analysis, coding the information and entering it into a separate record, and another GAO analyst verified the information for accuracy. All disagreements were resolved by a third GAO analyst. The analysts then tallied the total number of joint DOD doctrinal documents in which EOD functions were included. In addition, we reviewed and analyzed prior DOD reports' findings about doctrine and examined whether associated recommendations had been implemented. Specifically, we reviewed an EOD report from the Joint Staff[1] and an EOD report analyzing transforming the joint EOD force.[2] We also reviewed guidance on EOD activities from the services, including multi-service tactics, techniques, and procedures guidance. Additionally, we interviewed key EOD officials across DOD to ascertain the extent to which DOD has comprehensive joint EOD guidance and, if so, any potential benefits joint guidance has provided. Moreover, we discussed with DOD and service officials how EOD has been integrated jointly across DOD in areas such as joint operations and the joint training and equipping of EOD forces.

Finally, we met with EOD leadership and personnel from selected EOD units in all four military services and conducted 28 group discussions with EOD-qualified team members, team leaders, senior enlisted personnel, and officers to obtain their perspectives on issues related to military service-specific EOD missions, joint combat operations, training, equipment, and operational tempo. We used these data to provide illustrative examples throughout this report. A detailed discussion of how we conducted those group discussions follows, and more details of the themes from the group discussions can be found in appendix II. The group discussions included EOD-qualified personnel from the following military units:

U.S. Army

52nd Ordnance Group (EOD), Fort Campbell, Kentucky

- 49th EOD Company, Fort Campbell, Kentucky;
- 723rd EOD Company, Fort Campbell, Kentucky; and
- 788th EOD Company, Fort Campbell, Kentucky.

U.S. Navy

EOD Group One, Naval Amphibious Base Coronado, California

- EOD Mobile Unit One, Naval Base Point Loma, California;
- EOD Mobile Unit Three, Naval Amphibious Base Coronado, California; and
- EOD Mobile Unit Eleven, Imperial Beach, California.

U.S. Marine Corps

- 1st EOD Company, Camp Pendleton, California;
- 3rd EOD Company, Okinawa, Japan; and
- EOD Personnel Supporting the 3rd Marine Aircraft Wing, Marine Corps Air Station Miramar, California.

U.S. Air Force

- 96th EOD Flight Civil Engineering Squadron EOD Flight, Eglin Air Force Base, Florida.

We selected EOD units to visit based on information from the services for units that had recent deployment experience as well as the ability to provide sufficient quantities of EOD-qualified personnel who would be available to participate in our group discussions.

Our overall objective in using the group discussion approach was to obtain insight and perspectives from EOD personnel on training, equipment, operational tempo, joint military operations, and military service-specific responsibilities. Group discussions, which are similar in nature and intent to focus groups, involve structured small group discussions designed to obtain in-depth information about specific issues. The information obtained is such that it cannot easily be obtained from a set of individual interviews. From each location, we requested that each military service provide up to 10 volunteers to participate in our group discussions. We also conducted group discussions separated by rank and position. Specifically, we conducted separate group discussions that were comprised of officers, senior enlisted personnel, team leaders, and team members. At one location, two group discussions included both officers and senior enlisted personnel, and two other group discussions included all available EOD personnel assigned to those particular units. The number of participants per group discussion ranged from 2 to 12. Discussions were held in a semi-structured manner, led by a moderator who followed a standardized list of questions. The group discussions were documented by one or two analysts at each location. Group discussions were conducted between August 2012 and October 2012.

Composition of the Discussion Groups

We conducted 28 group discussions with EOD-qualified junior enlisted, noncommissioned officer, warrant officer, and commissioned officer personnel from the Army, Navy, Marine Corps, and Air Force. During each discussion, we asked participants to complete a voluntary questionnaire that provided us with supplemental information about each person's EOD background, including:

- Rank;
- EOD qualification level (Basic, Senior, or Master Badge level);
- Number of deployments to Iraq or Afghanistan;
- Support provided to or received from another military service; and
- Support to the Very Important Persons Protection Support Activity.

This information provided by participants helped us to ensure that we obtained a wide range of perspectives from qualified EOD personnel with a variety of EOD-related experiences. In total, we met with 188 EOD personnel. Table 1 below shows the composition of our various discussion groups.

Table 1. Composition of EOD Discussion Groups

	Army	Navy	Marine Corps	Air Force	Total
Total number of participants completing the questionnaire	39	53	67	29	188
Military Grade					
E-1 to E-3	5	0	0	2	7
E-4 to E-6	20	19	51	24	114
E-7 to E-9	7	16	9	3	35
W-1 to W-3	0	3	5	N/A[a]	8
W-4 to W-5	0	1	0	N/A[a]	1
O-1 to O-3	4	9	2	0	15
O-4 to O-6	3	5	0	0	8
No Response	0	0	0	0	0
EOD Qualification Level					
Basic EOD Badge	25	10	37	10	82
Senior EOD Badge	8	11	17	4	40
Master EOD Badge	6	30	13	15	64
No Response/Not Applicable	0	0	0	0	2[b]
Number of EOD Deployments in Support of Operation Iraqi Freedom					
None	28	19	39	7	93
1 to 5 deployments	11	34	28	21	94
6 to 10 deployments	0	0	0	0	0
11 or more deployments	0	0	0	1	1
No Response	0	0	0	0	0
Number of EOD Deployments in Support of Operation Enduring Freedom					
None	21	14	28	5	68
1 to 5 deployments	18	39	39	24	120
6 to 10 deployments	0	0	0	0	0
11 or more deployments	0	0	0	0	0
No Response	0	0	0	0	0
Received Support From/Gave Support to other Military Services					
Yes	26	46	50	25	147
No	13	7	17	4	41

Table 1. (Continued)

	Army	Navy	Marine Corps	Air Force	Total
Not Applicable	0	0	0	0	0
No Response	0	0	0	0	0
Participated in Very Important Persons Protection Support Activity Missions					
Yes	29	49	39	27	144
No	10	3	27	2	42
No Response	0	1	1	0	2

Source: GAO.

[a]N/A stands for Not Applicable. There are no warrant officers in the Air Force.

[b]Two participants identified their qualification level as "EOD Officer."

We performed content analysis of our group discussion sessions in order to identify the themes that emerged during the sessions and to summarize participant statements regarding EOD experiences and perceptions. Specifically, at the conclusion of all our group discussion sessions, we reviewed responses from the discussion groups and created a list of themes. We then reviewed the comments from each of the 28 group discussions and assigned comments to the appropriate themes. One GAO analyst conducted this analysis and a different GAO analyst checked the information for accuracy. Any discrepancies in the assignment of the comments to themes were resolved through discussion by the analysts. The information gathered during our group discussions with EOD personnel represents the responses of only the EOD enlisted and officer personnel present in our 28 group discussions and it is not projectable to other EOD personnel. We conducted this performance audit from May 2012 to April 2013 in accordance with generally accepted government auditing standards. Those standards require that we plan and perform the audit to obtain sufficient, appropriate evidence to provide a reasonable basis for our findings and conclusions based on our audit objectives. We believe that the evidence obtained provides a reasonable basis for our findings and conclusions based on our audit objectives.

Appendix II: Themes from the Group Discussions

In group discussions, we asked EOD personnel for their perspectives on what is working well and what needs improvement with regard to military service-specific missions, joint combat operations, training, equipment, and operational tempo. We analyzed participants' responses to identify the most common themes, which are summarized below.

EOD Organizational Alignment and Career Path: Some participants from each military service raised concerns about where EOD is aligned within their respective military service. Some also noted issues with career paths for EOD-qualified officers. For example, Army, Marine Corps, and Air Force personnel expressed concerns about EOD alignment under the Ordnance Corps (Army) or Engineers (Marine Corps and Air Force). They perceived these alignments as hampering EOD forces' influence for resourcing and operations. Additionally, Army officers reported that officer career paths are of concern because the EOD specialty is a technical area that is not the same as others in their organization. For example, the Army EOD officer career path is within the Logistics Branch of the Army, and all EOD officers are expected to have logistic skills; however, an Army officer noted that learning about managing fuel farms has nothing to do with EOD. Conversely, Navy participants viewed the officer career path as generally positive and noted that enabling an officer to stay within EOD for his or her entire career was beneficial to the EOD force.

Training: Participants in the majority of the group discussions said they felt positive about the training they received, and almost all of our group discussion participants reported that they wanted more training or more time for training. Some group discussion participants reported concerns about not being able to train with the same types of equipment, such as robots, as they would be using when deployed. Both Army and Marine Corps personnel expressed the desire for more training in homemade explosives and casualty care. Additionally, Army and Navy personnel reported that they needed additional access to training ranges. Army participants at Fort Campbell noted that access to training ranges is difficult to provide because there are not enough training ranges for all EOD companies to train at their installation. Likewise, Navy personnel in the San Diego area noted that they have issues accessing training ranges, particularly for exercises involving demolition.

Frequency of deployments: Participants in our discussions expressed mixed opinions regarding operational tempo. Some EOD personnel felt that the high pace of deployments and other missions put stress on EOD personnel and their families. For example, some personnel felt that EOD team leaders are getting burned out because they are always away from home. In addition, some personnel said that to cover missions clearing training ranges they had to turn down annual leave or cancel doctors' appointments because of staffing shortages. Also, some personnel noted that during their 12 months home between overseas deployments they are often away from their families because of the need to attend training or sometimes to travel in support of the Very Important Persons Protection Support Activity. Some personnel liked a high operational tempo and voiced concerns about having too much down time as operational tempo slows.

Support to Very Important Persons Protection Support Activity: Participants in many of our group discussions noted that the time they spent in support of the Very Important Persons Protection Support Activity exacerbated their high operational tempo, and some raised the concern that these missions are not an effective use of EOD skills. EOD personnel reported that these missions had a negative effect on EOD personnel and their families by taking them away from home too often. Further, they noted that to support these missions they missed training for overseas missions. Moreover, some participants said that the Very Important Persons Protection Support Activity missions were not a good use of the EOD skill set or did not use EOD to its full potential because EOD personnel are asked only to identify potentially explosive hazards but are not allowed to disarm anything that is found.

Funding for Training EOD Units: Some participants in our group discussions reported concerns about the adequacy of funding for EOD training. For example, some Army EOD personnel said that they sometimes had to buy materials needed for training aides, such as electrical tape and electronic parts. Moreover, Army and Marine Corps personnel reported that specialized training, such as post-blast analysis or homemade explosive courses, is expensive, which limits the number of people who can take the training. In particular, Marine Corps personnel expressed concerns that money for training might be scaled back in the future.

Incentive and Special Duty Assignment Pays: Participants in our discussion groups noted that incentive pay, special duty assignment pay, and retention bonuses are important to EOD personnel, and that availability of such pays is a factor in retaining EOD personnel, but that availability of incentives varied across the military services. Some participants from the Marine Corps raised issues regarding incentive pay, special duty assignment pay, and retention bonuses. Marine Corps personnel stated that, unlike the other military services, the Marine Corps does not provide any additional pays or bonuses for EOD, and they felt that not receiving these types of pay like the other services constituted an issue of equity.

End Notes

[1] We have previously testified that IEDs accounted for almost 40 percent of the attacks on coalition forces in Iraq during 2008. See Warfighter Support: Challenges Confronting DOD's Ability to Coordinate and Oversee Its Counter-Improvised Explosive Devices Efforts, GAO-10-186T (Washington, D.C.: Oct. 29, 2009). Also, we have previously reported that approximately 16,500 IEDs were detonated or discovered being used against U.S. forces in Afghanistan in 2011. See Counter-Improvised Explosive Devices: Multiple DOD Organizations Are Developing Numerous Initiatives, GAO-12-861R (Washington, D.C.: Aug. 1, 2012).

[2] See H.R. Rep. No. 112-78, at 116 (2011).

[3] See id. at 117.

[4] The committee directed DOD to establish a new consolidated budget justification display that fully identifies the services' baseline EOD budgets and encompasses all programs and activities of the EOD force for the functions of procurement; operation and maintenance; and research, development, testing, and evaluation. See id. at 116.

[5] For the purposes of this report, we obtained data only for active duty EOD forces. These forces comprise the majority of EOD personnel.

[6] Enlisted personnel in the Army, Navy, and Air Force may volunteer for the EOD force during basic training. In the Marine Corps, enlisted personnel must have achieved an enlisted rank 4. Officer personnel in the Air Force and Army can volunteer to become EOD-qualified but are part of the Civil Engineer and Logistics career fields, respectively. In the Navy, officer personnel can join the EOD career field upon becoming a commissioned officer. In the Marine Corps, enlisted personnel who are EOD-qualified can become warrant or limited duty officers.

[7] The Navy Expeditionary Combat Command mans, trains, and equips rapidly deployable and agile expeditionary forces to support warfare commanders' requirements for maritime security operations around the globe.

[8] See Department of Defense Directive 5160.62, Single Manager Responsibility for Military Explosive Ordnance Disposal Technology and Training (EODT&T), encl. 2, para. 3.a (June 3, 2011)(hereinafter cited as DODD 5160.62 (June 3, 2011)); see also Office of the Chief of Naval Operations Instruction 8027.6F, Naval Responsibilities for Explosive Ordnance Disposal, encl. 12 (Dec. 7, 2012).

[9] Common-type training is defined as training in EOD procedures required by two or more military services in the normal execution of their assigned missions. See DODD 5160.62, Glossary (June 3, 2011).

[10] The Army and Air Force calculations presented above do not reflect data on authorized EOD officers prior to 2005 and 2008, respectively, because service officials could not provide these data.

[11] The Very Important Persons Protection Support Activity provides EOD support to the U.S. Secret Service and the Department of State for the protection of high ranking government officials such as the President, Vice President, Secretary of State, and foreign heads of state visiting the United States. The services generally provide two-person EOD teams to search venues for explosive ordnance, among other duties.

[12] Department of Defense: Sustaining U.S. Global Leadership: Priorities for 21St Century Defense. January 2012.

[13] See H.R. Rep. No. 112-78, at 116 (2011).

[14] GAO, Standards for Internal Control in the Federal Government, GAO/AIMD 00-21.3.1 (Washington, D.C.: November 1999).

[15] The mission of the Joint IED Defeat Organization is to lead, advocate, and coordinate all DOD actions in support of the combatant commanders' and their respective joint task forces' efforts to defeat IEDs as weapons of strategic influence. A primary role for the organization is to provide funding and assistance to rapidly develop, acquire, and field counter-IED solutions.

[16] See Army Regulation 75-15, Policy for Explosive Ordnance Disposal (Feb. 22, 2005).

[17] Program element codes are the building blocks of the defense programming and budgeting system, and can be aggregated to display total resources assigned to specific programs, or specific military services, or in other ways, for analytical purposes.

[18] In this report, we use the term "authorized" to refer to manning levels for the EOD mission, as identified by the services.

[19] We have previously reported that the Army and the Marine Corps quickly acquired and fielded equipment to meet evolving threats. The equipment was initially supported with overseas contingency operations funding rather than through the Army's and Marine Corps' regular budgets and was not listed on unit authorization documents (modified tables of organization and equipment for the Army and tables of equipment for the Marine Corps). See Force Structure: Army and Marine Corps Efforts to Review Nonstandard Equipment for Future Usefulness, GAO-12-532R (Washington, D.C.: May, 2012).

[20] See Multi-Service Tactics, Techniques, and Procedures for Explosive Ordnance Disposal, ATTP 4-32.16(FM 4-30.16) /MCRP 3-17.2C/ NTTP 3-02.5/ AFTTP 3-2.32 (Sept. 2011).

[21] See Chairman of the Joint Chiefs of Staff Instruction 5120.02C, Joint Doctrine Development System, encl. A, para. 1.c (Jan. 13, 2012).

[22] The Air Land Sea Application Center is a multi-service organization established by the services' doctrine centers to develop tactical-level solutions to multi-service interoperability issues consistent with joint and service doctrine.

[23] This guidance describes each military service's EOD organizations, capabilities, equipment, doctrine, and training and provides joint EOD command and control options, as well as information for planning and conducting EOD operations in a joint environment. This guidance was updated in 2005 and 2011 to reflect EOD's evolving role in military operations.

[24] Department of Defense, Final Report of Assessment for Joint Explosive Ordnance Disposal. (Washington, D.C.: July 31, 2006).

[25] Squires & Fulcher, LLC. Joint Explosive Ordnance Disposal Force Transformation Analysis. A report prepared for the Department of Defense. May 30, 2007.

[26] Department of Defense, Final Report of Assessment for Joint Explosive Ordnance Disposal. (Washington, D.C.: July 31, 2006).

[27] See Chairman of the Joint Chiefs of Staff Manual 5120.01, Joint Doctrine Development Process, encl. B, para. 2 (Jan. 13, 2012).

[28] Id. para. 5.a.

End Notes for Appendix I

[1] Department of Defense, Final Report of Assessment for Joint Explosive Ordnance Disposal. (Washington, D.C.: July 31, 2006).

[2] Squires & Fulcher, LLC. Joint Explosive Ordnance Disposal Force Transformation Analysis. A report prepared for the Department of Defense. May 30, 2007.

In: Explosive Ordnance Disposal
Editor: Alain Michaux ISBN 978-1-62808-357-6

Chapter 2

EXPLOSIVE ORDNANCE DISPOSAL OPERATIONS*

Headquarters, Department of the Army

PREFACE

The Army Explosive Ordnance Disposal program can trace its roots back to April 1941. The United States was not yet at war, but was actively preparing for that eventuality. Embassy personnel and military observers were reporting on the actions of warring nations and as these reports were evaluated by the War Department, Intelligence Sections, recommendations were made concerning actions that should be taken by the United States. Delayed-explosion bombs were creating havoc in Europe, taking a heavy toll on lives and industry. It was expected that if the United States entered the war, we would experience bombing of our cities and industries. As a result, the need for a bomb disposal program in this country received immediate attention. It was decided that both military and civilian bomb disposal personnel would be trained by the Army. All responsibility for bomb disposal was placed under the U.S. Army Ordnance Department.

Army Tactics, Techniques, and Procedures (ATTP) 4-32 provides the doctrinal guidance for commanders, staffs, and leaders at all levels responsible for explosive ordnance disposal (EOD) operations. The doctrine in this ATTP

* This Army Tactics, Techniques, and Procedures publication, ATTP 4-32, was released December 2011.

is based on suggestions, insights, and observations developed from past and current operations and an analysis of the future modular force support in line with national military strategy. It provides staff planning, responsibilities, capabilities, and requirements relevant to the EOD mission.

This manual focuses on the employment of EOD forces. This ATTP explains the mission and organization of EOD units. It describes why EOD forces are an important part of the combined arms team throughout all phases of full spectrum operations. It also prescribes responsibilities for EOD forces in support of the Army, Joint Force, and subordinate commanders. The manual reflects and supports the Army operations doctrine as stated in Army Doctrine Publication (ADP 3-0).

This ATTP provides commanders and staff with doctrine for EOD operations and integration of EOD forces. It describes the EOD functional relationships which integrate EOD with the warfighting functions. This ATTP serves to illustrate Army EOD force structure operating as an enabler in the joint operating environment.

This publication applies to the Active Army, the Army National Guard (ARNG)/Army National Guard of the United States (ARNGUS), and the United States Army Reserve (USAR) unless otherwise stated.

The proponent of this publication is the United States Army Training and Doctrine Command (TRADOC). Send comments and recommendations on DA Form 2028 (Recommended Changes to Publications and Blank Forms [or in DA Form 2028 format]) directly to Commander, Combined Arms Support Command, Directorate of Training and Doctrine, ATTN: ATCL-O, 2221 Adams Avenue, Fort Lee, VA 23801-1809.

SECTION 1.EOD IN THE OPERATIONAL ENVIRONMENT

Army EOD supports the Army mission of prompt, sustained land dominance throughout full spectrum operations by identifying, diagnosing, rendering safe, exploiting, and disposing of all explosive ordnance or comparable threats regardless of filler; as required provides support to joint, interagency and multinational missions.

EOD units are the Army's preeminent tactical and technical explosives experts. They are properly trained, equipped, and integrated to attack, defeat, and exploit explosive ordnance. Explosive ordnance is defined in Joint doctrine as all munitions containing explosives, nuclear fission or fusion materials, and biological and chemical agents. This includes bombs and warheads; guided and ballistic missiles; artillery, mortar, rocket, and small arms ammunition; all mines, torpedoes, and depth charges; demolition charges; pyrotechnics; clusters and dispensers; cartridge and propellant actuated devices; electro-explosive devices; clandestine and improvised explosive devices; and all similar or related items or components explosive in nature. This also encompasses homemade explosives, improvised explosives, and explosive precursors. EOD units are made up of Soldiers that have the best tactical and technical training that Army and civilian academia can provide, to include Naval School EOD, Federal Bureau of Investigation Post Blast, and several others.

This section discusses employment of EOD in the operational environment (OE). It discusses how the EOD mission supports full spectrum operations in an era of persistent conflict. The threats range from an adaptive enemy using low technology and asymmetric methods to larger modernized forces able to engage deployed U.S. forces in more conventional, symmetrical ways. For the indefinite future, violent extremist groups, with or without state sponsorship, will continue to foment instability and challenge U.S. and allied interests. Our enemies are adaptive and will develop systems and tactics that exploit our vulnerabilities. They will continue to utilize a variety of explosive ordnance to include improvised explosive devices (IED). We must assume that the IED threat will evolve and persist even as better countermeasures are developed. In addition, future enemies will seek weapons of mass destruction and ways to employ them.

EOD in Full Spectrum Operations

1-1. Operations conducted outside the United States and its territories simultaneously combine three elements—offense, defense, and stability. Army forces use offensive and defensive operations to defeat the enemy on land. They simultaneously execute stability operations to interact with the populace and civil authorities. Civil support is the fourth element of full spectrum operations that occurs within the United States and its territories. EOD is an essential component to all phases within full spectrum operations.

1-2. EOD is a key asset in the protection of military and civilian personnel, critical assets, infrastructure, and public safety. Explosive ordnance threats may be present during all phases of full spectrum operations. In order to manage risk at the lowest possible level, commanders must integrate EOD throughout all phases of full spectrum operations.

Offensive and Defensive Operations

1-3. Offensive operations are combat operations conducted to defeat and destroy enemy forces and seize terrain, resources, and population centers. They impose the commander's will on the enemy (ADP 3-0). EOD supports the purposes of offensive operations by depriving the enemy of resources. This is accomplished through the disposal of captured enemy ammunition. EOD also supports the development of intelligence by performing post blast analysis, fragmentation analysis, site exploitation and producing weapons technical intelligence. EOD is the only asset capable of providing rendered-safe weapons in support of the weapons technical intelligence process and the larger counter-IED process. In addition, EOD supports the creation of a secure environment for stability operations by negating the enemy's ability to effectively use explosive ordnance as well as rendering safe and disposing of all unexploded explosive ordnance (UXO), both friendly and enemy.

1-4. Defensive operations are combat operations conducted to defeat an enemy attack, gain time, economize forces, and develop conditions favorable for offensive or stability operations (ADP 3-0). EOD supports the purposes of defensive operations by protecting the populace, critical assets and infrastructure. This is accomplished through the rendering safe and disposal of UXO, IED, homemade explosives and precursors, chemical and biological ordnance, nuclear ordnance, and weapons of mass destruction. Driving factors in the execution of the EOD mission during defensive operations include the level of security of the OE and the enemy's weapons of choice and tactics, techniques and procedures (TTPs). EOD units will focus on protecting joint, multinational, and host nation forces from all explosive threats. EOD supports the development of intelligence during defensive operations in the same way that it does during offensive operations. Additionally, EOD forces provide significant support to vulnerability and threat assessments.

Stability Operations

1-5. Stability operations encompass various military missions, tasks, and activities conducted outside the United States in coordination with other instruments of national power to maintain or reestablish a safe and secure environment, provide essential governmental services, emergency infrastructure reconstruction, and humanitarian relief (JP 3-0). The support provided by EOD during offensive and defensive operations continues during stability operations. The first aim is to limit adversaries' influence and isolate the populace from the enemy by establishing civil security. If the enemy poses a significant military threat, forces engaged in stability operations defend themselves and the populace from attacks. EOD units will focus on the safety and security of the forces, civilian populace, and infrastructure in regard to explosive threats. It is critical that commanders maximize the use of EOD qualified personnel in order to minimize the likelihood of collateral damage. Security is the foremost condition; it underpins all other sectors. During this phase EOD also focuses on the training, advising, and assisting of host nation and multinational forces in explosive ordnance recognition and response.

Civil Support Operations

1-6. Civil Support is Department of Defense (DOD) support to US civil authorities for domestic emergencies, and for designated law enforcement and other activities (JP 1-02). Civil support includes operations addressing the consequences of natural or man-made disasters, accidents, terrorist attacks, and incidents in the U.S. and its territories. Army forces conduct civil support operations when the size and scope of events exceed the capabilities of domestic civilian agencies.

1-7. Homeland Security Presidential Directive 19, Combating Terrorist Use of Explosives in the United States, states that the U.S. will counter the threat of explosive attacks aggressively by coordinating Federal, State, local, territorial, and tribal government efforts and collaborating with the owners and operators of critical infrastructure and key resources to deter, prevent, detect, protect against, and respond to explosive attacks.

1-8. EOD executes civil support operations by protecting civilian population, infrastructure, and property from the threat posed by military and improvised munitions. EOD responds, renders safe, exploits and disposes of all categories of explosive ordnance whether in a deployed area of operations (AO), on an Army installation or in support of civil authorities. These operations include conducting EOD procedures in response to requests for assistance from Federal, State, local, tribal and territorial authorities under immediate response authority and as directed by the Secretary of Defense. These EOD specific procedures include; identification, render safe procedures (RSP), recovery, and disposal of explosive ordnance. EOD is designated a first responder by federal statute and DOD directives. For more information on

EOD's role in civil support operations, refer to Appendix B, United States Operational Guidelines.

1-9. EOD supports the U.S. Secret Service (USSS) and Department of State (DOS) in protecting the President and Vice President, of the United States, designated foreign heads of state, and other dignitaries. The United States Northern Command (USNORTHCOM) is the supported Combatant Commander (CCDR) for EOD support to Very Important Persons (VIP) missions. Joint EOD VIP Protection Support Activity (JEODVIPPSA) is the USNORTHCOM action agent and is responsible for tasking of DOD EOD units for worldwide VIP mission support. For more information refer to Appendix C, EOD support to US Secret Service.

1-10. There are multiple policies and regulations regarding EOD support to civil authorities. These policies and regulations are outlined below:

- 40 Code of Federal Regulations, Part 260-265, Military Munitions Rule, 12 August 1997.
- DOD Directive 3025.13, Employment of DOD Capabilities in Support of the U.S. Secret Service (USSS), Department of Homeland Security (DHS), 8 October, 2010.
- DOD Directive 3025.18, Defense Support of Civil Authorities (DSCA), 29 December 2010.
- DOD Instruction 6055.17, DOD Installation Emergency Management (IEM) Program, 12 January 2009 (and Change 1, 1 November 2010).
- DOD 3150.8-M Nuclear Weapon Accident Response Procedures (NARP) 22 February 2005.

Joint and Multinational Operations

1-11. Army EOD command elements provide mission command to the joint force commander (JFC). Multiservice and multinational EOD units in support of land warfare operations are typically under the operational control of a single Army EOD command, due to common understanding of terminology, training, and equipment.

1-12. Army operations overseas will be conducted as part of a joint force which can be part of a larger multinational force. Joint service EOD formations may be utilized to augment formations where the need is greater or different than the capabilities of a single service. Supported units may be required to provide additional administrative and sustainment support to joint service EOD formations due to a limited capability to perform these functions internally. Integrating EOD with the staff ensures unity of effort, planning, preparation, execution, and assessment of joint EOD operations.

1-13. Depending on location, the commander may be employing multinational EOD forces. Army EOD assists multinational forces in the institutionalization of their EOD program. This assistance is accomplished through a partnering or exchange program with other nations. The Army may provide a qualified EOD noncommissioned officer (NCO) or commissioned officer for the development of training and mentoring programs. Joint and multinational EOD operations are further explained in ATTP 4-32.16, Multi-Service Tactics, Techniques, and Procedures for Explosive Ordnance Disposal in a Joint Environment.

Counter-Improvised Explosive Device Operations

1-14. Designing a successful counter-IED operation is a complex task that involves all echelons of the joint force and is based on a framework designed to assure the freedom of movement of friendly forces and enable commanders and staffs to plan and take proactive measures to identify and defeat IED events before they are successfully employed. Within each level of war there are key IED activities that influence operational planning. These activities must be viewed both individually and in the context of their relationship to the other activities that enable a counter-IED effort.

1-15. Counter-IED operations must take a holistic approach that incorporates intelligence, information, training, operations, materiel, technology, policy, and resourcing solutions designed to address all of the fundamentals of assured mobility, to include prediction, detection, prevention, neutralization, and mitigation. To a larger extent this approach should be considered in terms of joint interdiction which encompasses assured mobility as well as many other factors of warfare. Counter-IED operations are conducted across the phases of a military operation (shape, deter, seize the initiative, dominate, stabilize, and enable civil authority) and should be executed within multiple lines of operation, each to commence conditionally, and then to continue in parallel throughout a campaign. The three C-IED lines of operations are—Attack the Network: Joint force attack the network actions prevent the emplacement of the IED by attacking adversary vulnerabilities at multiple points. Key vulnerabilities within the adversaries IED employment system include: abilities to influence the support of the local populace, employment of IED TTP; the ability to maintain an IED component supply and distribution chain; the ability to establish and modify IED build-emplacement process.

Defeat the Device: The goal when defeating an IED is to prevent or mitigate its physical effects while marginalizing or preventing the adversary from exploiting the value of its effects to include propaganda and fear. Counter-IED defeat actions begin once the device has been emplaced and include detection, disarming it safely, and conducting a thorough forensics analysis.

Train the Force: JFCs must ensure the force is adequately trained prior to deployment. Areas of special interest include the development of relevant and current IED related TTP, drills, and standard operating procedures (SOP). Training should be designed to enhance individual and unit protection and the unit's ability to effectively operate in a high-threat IED environment. Training should also include those activities that facilitate the establishment and growth of coalition and partner nation IED defeat capabilities, including the transfer of counter-IED technology and U.S. force TTP.

1-16. EOD Groups execute command and control (C2) of counter-IED assets and operations in support of the JFC in their respective AOs. EOD Battalions can execute this mission for lower levels of allocation or during late phase stabilization operations, based on volume of missions, and span of control. It should be noted that late phase stabilization will mainly focus on building host-nation capacity and training the force. At this point, attack the network and defeat the device will be secondary concerns outside of force protection.

Summary

1-17. Army EOD plays an essential role throughout all phases of full spectrum operations. While in a deployed environment EOD supports the CCDR by providing the capability to identify, render-safe, collect, and exploit explosive ordnance; provide a tactical analysis of enemy IED TTPs; execute training and mentoring of joint, host nation and multinational forces; provide C2 for all phases of Counter-IED operations and EOD forces in the area of operations. Domestically, Army EOD provides immediate response to requests for assistance from civil authorities, executes installation force protection, and provides ordnance lifecycle support.

SECTION 2. EOD SUPPORT TO OPERATIONS

The military may find itself operating in several environments simultaneously. The challenge for the EOD commander is to ensure their assets are at the right place at the right time to support the commander in the AO. It is also important that the commander understands how to leverage the specialized capabilities EOD provides when seeking to maximize protection, while minimizing risk due to collateral damage.

EOD Functions

2-1. EOD functions are defined as categories of related EOD capabilities and activities grouped together to assist commanders at all echelons to effectively integrate and synchronize EOD forces. EOD supports the Commander's mission through its four functional areas; render safe, tactical and technical intelligence,

protection support, and disposal. The basic four EOD functions also directly support the Army's warfighting functions.

Render Safe

2-2. Render safe procedures are the portion of EOD operations involving the application of special EOD methods and tools to provide for the interruption of functions or separation of essential components of UXO to prevent an unacceptable detonation. (JP 1-02)

2-3. There is no "safe" procedure for disposing of UXO, IED, chemical, biological, radiological, and nuclear (CBRN) devices or other explosives, merely a procedure that is considered the least dangerous. Because maximum safety in any EOD operation can only be achieved through strict adherence to applicable safety precautions, protective measures used during the application of EOD procedures will be based on the possibility of a high-order detonation and based on operational considerations. This capability is reserved for only those personnel who are trained, certified and assigned to an EOD unit. EOD personnel are trained to perform EOD procedures on all explosive ordnance, including military munitions, landmines, IEDs, homemade explosives, weapons of mass destruction (WMD) and all other items which produce an explosive threat to operations.

EOD technicians are the only personnel equipped, manned, and trained to perform render safe procedures.

Tactical and Technical Intelligence

2-4. EOD maximizes combat effectiveness and facilitates timely situational understanding by collecting, processing, exploiting and analyzing components associated with explosive ordnance tactical and technical intelligence. Technical intelligence is derived from exploitation of foreign material, produced for strategic, operational, and tactical level commanders. The collection process begins at the incident site by conducting RSP, site exploitation, or a post-blast analysis. EOD determines the fuzing, firing, and functioning of explosive ordnance, collects forensic and biometric laden materials, and makes an analysis of the employment and emplacement techniques. EOD will then develop and integrate intelligence reports, assist in RSP development or modification, and prepare ordnance items for evacuation to organizations that have additional exploitation capabilities. Intelligence collected on explosive ordnance supports counterIED efforts by targeting, countering enemy TTP, assisting in development and refinement of electronic warfare countermeasures or exploitation of the electromagnetic spectrum associated with explosive ordnance.

Protection Support

2-5. Upon request from federal agencies, EOD will provide support to national special security events and counterdrug operations. Army EOD units provide force protection support to all Army facilities and installations, acts as the installation commanders principle advisor on explosive threats and is a DOD first responder.

Disposal

2-6. EOD personnel perform disposal procedures throughout all phases of full spectrum operations. Disposal operations are generally categorized as the final disposition of explosive ordnance by EOD personnel, which may include disposal by detonation or by a controlled burn.

Render Safe	Tactical and Technical Intelligence	Protection Support	Disposal
-UXO -IED -CEA -ERW -CBRN -WMD -SOF Support -Mortuary Affairs -JPAC Support	-UXO -IED -CEA -ERW -Fragmentation Analysis -First Seen Ordnance -Site Exploitation -Weapons Technical Intelligence	-Provide Training -HMA -NAIRA -Support to Civil Authorities -Vulnerability Assessment	-UXO -IED -CEA -ERW -Range Clearance -Stuck Rounds -Misfired Munitions -Battle Damaged Weapons Systems -CBRN -Depleted Uranium -SOF Support -Mortuary Affairs -JPAC Support -Unserviceable Ammo

* Description of capabilities can be found in EOD Organization and Capabilities.

Figure 2-1. EOD Functions and Capabilities.

EOD within the Army Warfighting Functions

2-7. The Army warfighting functions provide commanders a common framework within which to link the EOD functions and their respective capabilities to the synchronized application of combined arms.

Mission Command

2-8. Mission command is the exercise of authority and direction by commanders, supported by their staffs, using the art of command and the science of control to integrate warfighting functions in the conduct of full spectrum operations. Mission command uses mission orders to ensure disciplined initiative within the commander's intent, enabling agile and adaptive commanders, leaders, and organizations (ADP 3-0).

2-9. The EOD company, battalion and group commands and staffs may exercise synchronized mission command of Army, joint, and multinational EOD forces. Senior EOD personnel can provide the expertise to plan, prepare, execute, assess and integrate external EOD formations into the supported unit.

Movement and Maneuver

2-10. In the face of adaptive challenges in an era of persistent conflict it is imperative for the Army to produce leaders and forces that exhibit a high degree of operational adaptability. Achieving the necessary level of operational adaptability requires the Army to build upon a foundation of two broad responsibilities within the framework of full spectrum operations; combined arms maneuver and wide area security.

2-11. Army forces conduct combined arms maneuver to gain physical, temporal, and psychological advantages over an enemy. Army forces establish wide area security to consolidate gains and ensure freedom of movement and action. Army forces employ combined arms maneuver and wide area security to seize, retain, and exploit the initiative. Army forces capable of effective combined arms maneuver and wide area security at both the operational and tactical levels provide joint force commanders the ability to deter conflict, prevail in war, and succeed in a wide range of contingencies.

2-12. EOD supports freedom of movement and maneuver most directly through the render safe and disposal of explosive ordnance threats which restrict or prevent the commander's ability to move forces.

2-13. The maneuver commander may order a tactical enabling mobility operation, such as clearing, to remove explosive ordnance in order to facilitate mobility on a critical route (route clearance) or area. Clearing requires the commander to remove all enemy forces and eliminate organized resistance in an assigned area (FM 3-90). The force does this by destroying, capturing, or forcing the withdrawal of enemy forces so that they cannot interfere with friendly unit missions. EOD forces may be utilized to support maneuver or engineer forces for clearing. The engineer or EOD Commander provides recommendations to the maneuver commander on organization, procedures, and support requirements for clearance operations.

2-14. Commanders may employ mobility operations in order to enable the IED defeat mission especially during stability operations. The successful utilization of the IED as a weapon of choice by the enemy is the final product of a network of enemy activities referred to as the IED network. The ability of EOD to render safe explosive ordnance, including IEDs, preserves forensic and technical intelligence which facilitates attacking those individuals or groups engaged in the IED network.

Sustainment

2-15. EOD assists commanders with identifying and advising on explosive ordnance. EOD can support and advise the commander on the safe location and establishment of ammunition supply points (ASP) and ammunition transfer and holding points. EOD can also assist and advise the commander on the safe storage, shipment and disposal of captured enemy ammunition (CEA), or ammunition that has been damaged or is unserviceable, to include explosive ordnance and bulk explosives. EOD assists commanders in the preservation of sustainment assets and lines of communication.

2-16. EOD provides the capability to remove stuck rounds and download misfired munitions from all weapon systems, both friendly and enemy, in order to resolve weapon system failures and maintain operational readiness. EOD will also inspect and clear battle damaged vehicles and aircraft prior to repair.

2-17. Mortuary affairs personnel often encounter explosive ordnance while recovering and processing remains. EOD support is crucial in order to ensure that all remains are clear of explosive ordnance before mortuary affairs conduct their mission.

Protection

2-18. The primary role of EOD is to protect life and property or priority assets. EOD provides commanders with the ability to quickly respond to explosive ordnance threat situations as they occur. The mission responsibility of EOD applies to both their assigned home station, as well as deployed operational environments. This responsibility encompasses not only EOD support to various installations and echelons of the Army, but also to federal, state, and local agencies within the United States. In a deployed AO, EOD provides embedded support to missions and operations where the commander requires immediate support against likely explosive ordnance threats.

2-19. EOD provides bomb threat and search procedures and instruction, facility site survey assistance for weapon storage sites, explosive safety assessments, and developing and implementing EOD emergency response, antiterrorism, and force protection plans. EOD's ability to render safe explosive ordnance is critical to the commander's ability to protect key facilities and infrastructures. Rendering safe explosive ordnance decreases the likelihood of high order detonation and the spread of contamination.

2-20. EOD provides the capability to plan, direct, and support range clearance operations by conducting identification and disposal of UXO to facilitate range maintenance and target insertion on worldwide military installations and operations bases.

2-21. EOD supports the Army counter-IED Strategy by assisting in the establishment of counter-IED situational training exercise lanes and search houses. EOD forces can also provide training in explosive ordnance recognition and reporting, bomb threat search procedures and evacuation, site vulnerability assessments, and unit SOP preparation and validation. This training increases the effectiveness of the commander's protection program.

2-22. With command direction and approval, Army EOD trains foreign military and law enforcement in explosive ordnance identification and limited RSP. EOD advises foreign nationals in support of humanitarian de-mining operations.

2-23. At the conclusion of an EOD response the EOD Team Leader may create a storyboard detailing the specific information that pertains to the incident. Storyboards include photos, cardinal directions, emplacement methods, type and size of explosive ordnance, functioning and firing of the device, and other pertinent information. These storyboards can be used by courts, both foreign and domestic, to aid in the prosecution of suspected bombers. EOD Soldiers may also be required to testify in courts as subject matter experts concerning EOD incidents.

Intelligence

2-24. As part of wide area security and combined arms maneuver, commanders employ intelligence collectors, analysts, and associated systems. The continuous interplay between the various intelligence disciplines and units conducting operations requires intelligence professionals and operators to collaborate at the lowest levels. This continuous dialogue creates timely, relevant, and clear information upon which commanders base their plans, decisions, and orders.

2-25. Co-creation of context is a continuous process in which commanders direct intelligence priorities to drive operations, and the intelligence that these operations produce causes commanders to refine operations based on an improved understanding of the situation. The corps headquarters synchronizes intelligence, surveillance, and reconnaissance (ISR) activities from the national to the tactical level to assist subordinate commanders in developing a common situational understanding of their AO.

2-26. The corps headquarters, EOD group, division headquarters and EOD battalion play an important role in developing the co-creation of context for tactical and technical intelligence of explosive ordnance. The division headquarters and EOD battalion may augment the brigade combat team (BCT), maneuver enhancement brigade (MEB) and subordinate units with an EOD company and EOD platoons to enable the collection of data and analysis of explosive ordnance.

2-27. EOD provides tactical and technical intelligence collection on UXO/IED/homemade explosive and precursors, IED components, and first seen ordnance during offensive, defensive and stability operations. It develops an understanding of enemy operations through IED employment and emplacement methods, IED designs and functions, and forensics which identify individuals and groups critical to enemy operations. EOD intelligence collection can positively identify specific persons of interest or high value individuals within an IED network; it can also provide details to establish profiles which identify unique signatures of a bomb maker. EOD personnel are responsible for ensuring that an area is explosively safe before allowing specialized intelligence enablers to conduct collection operations.

2-28. Render safe procedures preserve evidence for exploitation of tactical and technical intelligence. Explosive elimination of an IED frequently destroys and disperses material of intelligence value which leads to attacking the IED network. EOD RSPs for defeating the device directly provide targeting intelligence for attacking the network. EOD RSPs are the only techniques which link the defeat the device line of effort with the attack the network line of effort due to the intelligence collection for the targeting function.

2-29. The tactical and technical analysis of explosive ordnance contributes to the protection warfighting function by informing the force on explosive ordnance capabilities and enemy employment TTPs. EOD forces enable the co-creation of context on explosive ordnance by rapidly disseminating early warning to achieve the commanders' desired degree of protection for personnel and physical assets.

Fires

2-30. The fires warfighting function is the related set of tasks and systems that provide the collective and coordinated use of Army indirect fires, air and missile defense fires, electronic attack and joint fires capabilities, integrated through the targeting and defense design process. Targeting is the process of selecting targets and matching the appropriate response to them on the basis of operational requirements and capabilities. Material collected from IED sites and weapons caches by EOD personnel provide forensic data used to link identities to specific individuals within a group or IED network.

Eod Organizations and Capabilities

2-31. The six warfighting functions along with leadership and information compose the eight elements of combat power. Combat power is the way commanders conceptualize capabilities. Combined arms maneuver and wide area security apply the elements of combat power. Commanders and EOD leaders synchronize EOD functions and capabilities within the warfighting functions to support combined arms maneuver and wide area security. Commanders at all echelons must understand EOD capabilities and limitations to effectively utilize EOD forces. Below is a discussion of EOD Organizations and capabilities.

Group

2-32. The EOD Group is a functional C2 headquarters for Army EOD and if directed by the CCDR, is capable of conducting C2 for EOD missions of two to six EOD battalions. If directed by a JFC, the EOD Group Commander may be designated as the commander of a Joint EOD Task Force (JEODTF) or Counter-Improvised Explosive Device Task Force.

2-33. The EOD Group may be organized under a theater army, corps, joint task force (JTF), or combined joint task force (CJTF). In support of a specific operation, operation order (OPORD), operation plan (OPLAN), or concept plan (CONPLAN), the EOD Group in coordination with the supported higher headquarters, will conduct a mission analysis to determine the capabilities, task organization, and command and support relationships necessary to accomplish the mission.

2-34. Planners at all echelons must carefully consider the appropriate relationship needed for each situation to balance responsive support with the flexibility to distribute the low-density and high-demand EOD forces across the AO. The EOD Group is designed to perform EOD operations as required by the CCDR or JFC for the echelon it is supporting when the number of EOD units or the synchronization of EOD capabilities requires brigade level C2 capability.

2-35. If directed by the CCDR or JFC, it synchronizes EOD support across multiple organizations that control an AO. Typically this would be in support of a corps, JTF, or multinational headquarters. The EOD Group coordinates with the higher headquarters to plan, prepare, and assess EOD operations and provide training readiness oversight. It conducts analysis of explosive ordnance threats and associated trends; develops solutions; implements and coordinates measures to counter explosive threats.

2-36. The EOD Group interfaces, coordinates, and integrates with the corps, JTF, or multinational headquarters and other organizations contributing to explosive ordnance solutions, which include tactics, material solutions, and weapons technical intelligence. As required by the CCDR or JFC, it conducts mission analysis to analyze the explosive ordnance threat and EOD force requirements.

2-37. The EOD Group provides recommendations to the CCDR or JFC on apportioning and allocating EOD battalions and EOD companies required to counter the explosive ordnance threat. The EOD Group has limited organic structure and must rely upon other organizations for additional capabilities. The EOD Group and EOD battalions establish and operate EOD incident and technical intelligence reporting systems. They coordinate; monitor status; and as required expedite requests for special EOD tools, EOD specific equipment, and demolition materials.

2-38. The EOD Group in coordination with higher headquarters can constantly assess operations and the evolving explosive ordnance threat. The EOD Group can provide higher headquarters with assessments which require immediate adjustments of EOD units to attack the threat. It can provide critical and time sensitive adjustments to specific EOD TTPs as well as Army TTPs. A direct chain of command allows EOD higher headquarters to mass forces at a time and place in support of the commands' mission. This structure allows the group to rapidly adjust TTPs within the current operational environment. Refer to Appendix A for the EOD Group structure.

Battalion

2-39. The EOD Battalion is a functional C2 headquarters for Army EOD and if directed by the CCDR or JFC is capable of providing EOD mission command; planning, integrating, coordinating and directing the execution of EOD missions for three to seven EOD companies to include joint EOD forces.

2-40. The EOD Battalion or battalion minus may deploy as the senior mission command element for EOD operations and may be organized under a theater army, corps, division, CJTF or JTF. The EOD Battalion may deploy with its assigned EOD Group or may have operational control (OPCON) designated to a separate EOD Group.

2-41. In support of a specific operation, OPORD, OPLAN, or CONPLAN, the EOD Battalion in coordination with the supported higher headquarters, will conduct a mission analysis to determine the capabilities, task organization, and command and support relationships necessary to accomplish the mission. This may include tactical and weapons technical intelligence; acquisition; management; administrative and limited logistical support for three to seven EOD companies operating in a division or JTF AO.

2-42. The EOD Battalion may conduct planning, preparation, and assessment of division, CJTF or JTF EOD operations; provide recommendations on the apportionment and allocation of EOD forces; and provide training readiness oversight to all subordinate EOD forces.

2-43. The EOD Battalion provides liaison teams to augment planning, preparation, execution and assessment of EOD operations to the supported division, CJTF or JTF. The EOD battalion commander will coordinate and make recommendations to higher headquarters for synchronizing EOD operations.

2-44. Explosive ordnance material and information used to produce tactical intelligence and weapons technical intelligence are analyzed and the results are disseminated by the EOD Battalion to continually monitor changes in threat TTPs. The result of this analysis is provided to EOD companies and higher headquarters for inclusion into their operating procedures.

2-45. The EOD Battalion will facilitate and monitor fielding of new EOD specific equipment to subordinate EOD companies. Staff coordination must be made at the appropriate division or JTF so that supporting unit requirements are included in sustainment planning. The S-4 coordinates support and the process for requesting support. This coordination includes all company operations officers and first sergeants who ensure logistics support is integrated into operations planning. Refer to Appendix A for the EOD Battalion structure.

Control Team

2-46. The EOD Control Team may provide C2 of 1-2 EOD companies in an AO when an EOD Battalion is not assigned to that AO. The EOD Control Team may be utilized when the EOD Company supports a nonEOD battalion. It will coordinate and provide recommendations as required to higher headquarters for apportioning, allocating, and synchronizing EOD operations. It may provide limited planning, preparation, and assessment for EOD operations in support of the Division, CJTF or JTF.

2-47. Functional or multifunctional brigades may require augmentation by an EOD Control Team to provide additional C2 as EOD companies have limited capabilities to integrate EOD operations.

Company

2-48. The EOD Company provides mission command, explosive ordnance related tactical intelligence, weapons technical intelligence, management, administrative support, and training readiness for 3-5 EOD platoons. These EOD companies are typically organized under a BCT, MEB, or EOD Battalion.

2-49. The EOD Company in DS of a brigade may be task organized within that brigade for the purposes of exchanging information and operational planning. The EOD Company Commander can, on a temporary basis, further task-organize EOD forces to support subordinate maneuver task forces, subordinate battalions and companies.

2-50. The EOD Company Commander must provide C2 for the EOD Company and simultaneously be the BDE CDR's counter-IED subject matter expert. The EOD Operations section should integrate with the supported unit S2 and S3 sections. This will maximize the effectiveness of EOD support and provide the BDE situational awareness of explosive ordnance threats in the AO. The EOD Company should be involved at all levels of operational planning where EOD forces are required. The EOD Company will provide input to staff products related to EOD operations and capabilities.

2-51. Coordination and integration of EOD operations at all echelons is critical due to the limited EOD force. In support of a specific operation, OPORD, OPLAN, or CONPLAN, the EOD Company in coordination with the supported higher headquarters, will conduct a mission analysis to determine the capabilities, task organization, and support relationships necessary to accomplish the mission. Refer to Appendix A for the EOD Company structure.

EOD Organization	Supported Organization	Modeling Rule of Allocation
III EOD	•Theater Army •Corps •JTF •CJTF	•1 per Theater Army •1 per Corps •1 per JTF •1 per CJTF •1 per 2-6 EOD Battalion
II EOD	•Theater Army •Corps •Division •JTF •CJTF	•1 per Theater Army •1 per Corps •1 per Division •1 per JTF •1 per CJTF •1 per 3-7 EOD Companies
I EOD	•BCT •MEB •EOD Battalion	•1 per BCT •1 per MEB •1 per 3-5 EOD Platoons

Figure 2-2. EOD Organizational Chart.

EOD Conus Support Company

2-52. The Continental U.S. (CONUS) Support Company is part of the EOD generating force and provides EOD capabilities to federal, state, and local agencies. It coordinates support to the US Secret Service for the protection of VIPs, to the Federal Bureau of Investigation for counter-terrorism activities, and various other agencies for Civil Defense and EOD response to the CONUS landmass. The existence of this company does not relieve EOD companies of maintaining an area of response, in accordance with AR 5-9 in order to execute immediate response support to civil authorities. For more information on EOD's role in civil support operations, refer to Appendix B, United States Operational Guidelines.

2-53. EOD CONUS Support Company should be utilized for national capital region area of support missions to the greatest extent possible. The Company coordinates responses to EOD incidents requiring more than one response team. The Company maintains the capability to respond to 120 incidents per 24 hours.

Eod Weapons of Mass Destruction Company

2-54. The EOD Weapons of Mass Destruction Company provides highly technical EOD operations and containment procedures for WMD in support of joint or interagency operations. It has the ability to respond anywhere in the world with two fully capable eight person platoons as part of the Joint Technical Operations Team to defeat or mitigate the effects of WMD against the Unites States. The unit has the capability to provide four WMD platoons to support the Army or other U.S. agency in support of missions to defeat or mitigate WMD directed against the U.S. or national interest.

Conduct Unexploded Explosive Ordnance Operations

2-55. Unexploded explosive ordnance is explosive ordnance that has been primed, fuzed, armed, or otherwise prepared for action and which has been fired, dropped, launched, projected, or placed in such a manner as to constitute a hazard to operations, installations, personnel, or materiel and remains unexploded either by malfunction or design or for any other cause. (JP 1-02)

2-56. It is the responsibility of leaders at all levels to ensure that personnel are trained to recognize and react to UXO threats. Failure to properly train personnel will cause injuries or death. Personnel training and OPORDs should include UXO identification, reporting, marking, and protective measures procedures. U.S. and foreign UXO pose a threat to personnel, facilities and infrastructure, equipment, and movement and maneuver. Understanding the effects of UXO maximizes the effective use of resources and minimizes the impacts to operations.

2-57. EOD enhances a commander's combat power by identifying, rendering safe, or disposing of UXO. EOD has the ability to conduct fragmentation analysis and technical intelligence on all UXO, to include new or first seen explosive ordnance.

2-58. When a Soldier identifies a possible UXO they should immediately stop and notify others of the potential danger. The area will be searched for other threats and a perimeter will be formed in order to maintain protection from the UXO and not allow anyone to enter the danger area. It is important to maintain frontal and overhead protection from the UXO while still maintaining situational awareness. An EOD 9-Line will be sent to the finding units' higher headquarters requesting EOD support. The finding unit will remain on site until the EOD team arrives on scene and clears the area of all explosive threats. If a unit finds a possible UXO but is unable to stop, the area should be marked and the location of the possible UXO reported. More detailed information on UXO can be found in ATTP 4-32.2, UXO Multi-Service Tactics, Techniques, and Procedures for Unexploded Ordnance.

Conduct Improvised Explosive Device Operations

2-59. An IED is a device placed or fabricated in an improvised manner incorporating destructive, lethal, noxious, pyrotechnic or incendiary chemicals and designed to destroy, incapacitate, harass, or distract. It may incorporate military stores, but is normally devised from nonmilitary components. (JP 1-02)

2-60. The proliferation of IEDs poses a pervasive threat as an effective weapon system. The demonstrated lethality of this threat has influenced unit operations, U.S. policy, and public perception and is expected to remain a major component of future conflicts.

2-61. EOD enhances a commanders' combat power by identifying, rendering safe, or disposing of IEDs. EOD has the ability to conduct site exploitation, post blast analysis, fragmentation analysis, and intelligence reporting in support of counter-IED efforts.

2-62. When a Soldier identifies a possible IED, they should immediately stop and notify others of the potential danger. The area will be searched for other threats and a perimeter will be formed in order to maintain protection from the IED and not allow anyone to enter the danger area. It is important to maintain frontal and overhead protection from the IED while still maintaining situational awareness. IEDs can contain chemical, biological, and radiological enhancements. Soldiers need to be aware of the potential for exposure to these enhancements and be prepared to act accordingly. An EOD 9-Line will be sent to their higher headquarters requesting EOD support. The finding unit will remain on site until the EOD team arrives on scene and clears the area of all explosive threats. If a unit finds a possible IED but is unable to stop, the area should be marked and the location of the possible IED reported.

5's AND 25's

Placing multiple IEDs in one location is a common enemy TTP. Soldiers need to thoroughly search the area for multiple devices. If in a vehicle, Soldiers should search a 5 meter area from inside the vehicle. Once dismounted the underside of the vehicle needs to be searched and a 25 meter search of the area should be conducted.

Provide Training

2-63. EOD can provide bomb threat and search training to military and civilian personnel. EOD does not normally respond until a suspected device has been located. Once a search team finds a suspected IED, they should evacuate the area and report the device to EOD. EOD Soldiers DO NOT search for suspected IEDs based on bomb threats. The primary reason for this is that the personnel most familiar with a facility or area are in the best position to determine when something is suspect, or out of place. Workers assigned to a specific facility will be extremely helpful to EOD personnel in narrowing the area of focus for a suspected or confirmed bomb or IED.

2-64. During stability operations the establishment of a bomb disposal program for the host nation is essential to ensure that they are able to take over the mission of disposing of explosive ordnance once U.S. forces turn over responsibility. EOD supports this training through partnership operations providing commissioned officers and NCOs to advise and assist host nation forces in explosive ordnance recognition, disposal, and limited RSPs.

2-65. A combat theater will have a large array of explosive ordnance threats that may be encountered. This threat is constantly changing and EOD personnel will have the most up to date knowledge on new TTPs related to explosive ordnance that the enemy is using. In order to ensure that joint and multinational forces are aware of these threats, EOD will establish training lanes in order to recreate explosive ordnance threats that are being encountered. EOD personnel build inert training devices from bomb making material that have been captured and place these devices on the training lane using the same methods as the enemy. EOD conducts classes for joint and multinational forces at the training lanes to give personnel the opportunity to gain a better understanding of the current TTPs being used and the possible effects of explosive ordnance.

Conduct Multiple Unexploded Explosive Ordnance Operations

2-66. Any situation where more than one UXO is found is a multiple ordnance incident. Multiple ordnance incidents may include, but are not limited to, destroyed or damaged ASP or ordnance caches, ammunition convoy accidents, improved conventional munitions, or airfield clearance operations.

2-67. Only EOD qualified personnel should conduct or directly supervise multiple ordnance clearance operations. No other DOD personnel are trained or equipped to conduct these operations. EOD personnel are responsible for the reconnaissance and positive identification of UXO, enforcement of downrange safety measures, and determination and conduct of render safe and disposal procedures.

2-68. Although some incidents involving multiple ordnance may be completed very quickly, multiple ordnance incidents may take several days or weeks or multiple EOD teams to complete, depending upon the magnitude (quantity of UXO) and the complexity of the situation. The general methodology for conducting multiple ordnance operations utilizes a four-phased approach, consisting of initial planning, hasty reconnaissance or immediate action, detailed reconnaissance, render safe and disposal. The senior EOD Team Leader on site plans and executes the operation. More detail on multiple UXO response can be found in FM 3-100.38 Multi-Service Tactics, Techniques, and Procedures (MTTP) for UXO Operations.

Conduct Captured Enemy Ammunition and Explosive Remnants of War Operations

2-69. Captured enemy ammunition is discovered or captured quantities of stored, cached, or abandoned enemy ammunition and explosives. Explosive remnants of war and captured enemy ammunition should be considered operationally similar. Generally, captured enemy ammunition is characterized as all ammunition products and components a hostile foreign force produced or procured for use in combat or offensive operations and that are now in the control or custody of a U.S. military force or DOD component. Captured enemy ammunition can also include North Atlantic Treaty Organization (NATO) or U.S. manufactured munitions that may not have been under continual U.S. custody or control. Items including, but not limited to, landmines, bulk explosives, and UXO are categorized as explosive remnants of war following conclusion of major combat operations. The distinction between captured enemy ammunition and explosive remnants of war typically exists as an operation shifts from offensive and defensive operations into stability operations.

2-70. Recovery and evacuation of captured enemy ammunition and explosive remnants of war are command responsibilities at all echelons. Caches of captured enemy ammunition may be as small as one or two pieces of ordnance or as large as several thousand pieces of ordnance. Caches can be found in a variety of locations, such as schools, homes, religious centers, hospitals, sewage systems, or buried in farmers' fields. Large quantities of captured enemy ammunition and explosive remnants of war may also be found in ASPs similar to U.S. Army ASPs. Basic captured enemy ammunition and explosive remnants of war planning and execution factors are the same regardless of quantity encountered. Commanders and leaders at all levels must involve EOD and trained ammunition handling specialists in the planning of captured enemy ammunition and explosive remnants of war operations and should utilize them in the execution of captured enemy ammunition operations. Refer to Service regulations and theater policy for guidance on disposal of captured enemy ammunition by other than EOD.

2-71. When explosive ordnance is found, the capturing unit should not enter the area. It should be assumed that the area has been booby trapped. The capturing unit will maintain a security perimeter and request EOD assistance. When EOD arrives on scene they will make the determination if the explosive ordnance is safe for transportation or if it needs to be destroyed on site. Only U.S. EOD personnel are authorized to dispose of captured enemy ammunition. The capturing unit is responsible for munitions security until disposition instructions are received.

The EOD Team Leader is responsible for determining all necessary explosive safety measures and will advise the capturing unit on how to apply these measures.

The capturing unit is responsible for assisting in safeguarding, recovering, and evacuating munitions.

Destruction of Captured Enemy Ammunition

Prior to the destruction of captured ammunition, EOD personnel must make a positive identification of the munitions type and filler; destroying ammunition when the contents are unknown can result in the possible release of CBRN contamination. It is not always possible to identify ordnance types by color codes or outside markings. Some countries do not have standard markings for ammunition.

If security is not maintained on the site, the site must be re-cleared for booby traps. EOD performs initial assessment of captured enemy ammunition sites, eliminates immediate hazards, conducts technical intelligence procedures on first-seen ordnance, and recommends disposition to capturing unit.

Support to Humanitarian Mine Actions

2-72. EOD provides personnel in support of the DOD Humanitarian Mine Action (HMA) program. The goal of the DOD HMA program is to relieve human suffering and the adverse affects of landmines and explosive remnants of war on noncombatants while advancing the combatant commanders' security cooperation strategies and U.S. national security objectives.

2-73. Under 10 USC 407(a)(3), members of the US Armed Forces while providing humanitarian demining assistance shall not "engage in the physical detection, lifting, or destroying of landmines (unless the member does so for the concurrent purpose of supporting a United States military operation)."

2-74. EOD is a priority force for HMA missions. EOD personnel that participate in a HMA mission must first attend the Humanitarian Demining Training Center course located at Fort Leonard Wood, Missouri. The Humanitarian Demining Training Centers primary mission is to train U.S. military personnel in accordance with International Mine Action Standards. More detailed information on HMA can be found in CJCSI 3207.1B, Military Support to Humanitarian Mine Actions.

SUPPORT SITE EXPLOITATION OPERATIONS

2-75. Site exploitation is systematically searching for and collecting information, material, and persons from a designated location and analyzing them to answer information requirements, facilitate subsequent operations, or support criminal prosecution. (ATTP 3-90.15) Site exploitation is an enduring and integral mission in support of full spectrum operations. EOD forces provide support in planning and executing site exploitation operations. Site exploitation operations focus on searching facilities, materiel, weapons, equipment, personnel, and infrastructures to detect, collect, process, and evaluate material, components, information, and forensic samples.

2-76. EOD conducts site exploitation and technical intelligence collection and analysis of explosive ordnance, and components. Exploitation operations may additionally include RSPs, and clearance of any booby traps, if necessary, to render explosive ordnance materiel, systems, and equipment ineffective for use against U.S. and multinational forces.

2-77. Site exploitation may be either preplanned or targets of opportunity. Because EOD support can be a limited commodity with high demand, tactical commanders must plan for EOD support, when possible. In cases when EOD cannot be integrated into the site exploitation team, senior tactical commanders plan for the movement of EOD teams to sites requiring their support.

CONDUCT WEAPONS TECHNICAL INTELLIGENCE COLLECTION AND EXPLOITATION

2-78. Weapons technical intelligence is the intelligence derived from the technical and forensic collection and exploitation of first seen ordnance items, IEDs, associated components, improvised weapons, homemade explosives and precursors, and other weapon systems. Weapons technical intelligence has the ability to provide immediate feedback to the commander regarding force protection, targeting, sourcing, and support to prosecution. It encompasses traditional technical intelligence as well as analysis of tactical design and intended outcomes, enabling pattern analysis and event signature development.

2-79. EOD is the proponent for technical intelligence on explosive ordnance to determine it's fuzing, firing, and functioning. EOD can also determine the type of filler that the explosive ordnance contains, such as military grade high explosives, homemade explosives or other enhancements. Explosive ordnance post-blast analysis, site exploitation, design analysis, function diagnostics, and forensics and biometrics collection by EOD allows for the prediction and prevention of IED emplacement; and the detection and targeting of individuals and groups within the IED network. Forensics or biometrics collection by EOD, such as the collection of materials that can provide latent fingerprints, assists with the identification of IED suspects and terrorists groups. These individuals and groups include the local support groups, orders group, suppliers, IED makers, and the emplacer(s).

2-80. The EOD team will generate an intelligence report such as a Complementary Technical Intelligence Report-Type B and forward all data and physical materiel to the EOD Battalion for further exploitation. The report may contain digital photographs, digital x-rays, digital copies of detailed drawings, measurements and a description of the item along with theoretical functioning of the device. Simultaneously, the EOD Company provides the supported unit with all necessary information to enable the target information center to perform pattern and predictive analysis. The EOD Battalion, in direct support of operational-level forces, will analyze the intelligence report and provide render safe procedures or adaptive TTPs to the EOD Company while transferring custody of weapons technical intelligence related materiel to the operational-level exploitation elements.

Support Range Clearance Operations

2-81. Range clearance operations are the elimination of explosive ordnance from weapons ranges to facilitate range maintenance or target emplacement. Range clearance operations are conducted by EOD personnel both in garrison and deployed. Range clearance operations require long-range planning due to their complexity. Each operation has specific requirements that vary for each range and installation. These operations are the responsibility of the installation. The installation is responsible for planning, funding, and providing support personnel and resources necessary to conduct range clearance operations. EOD forces are responsible for the detailed operational planning and safe execution of the operation. Render safe procedures, removal of explosive ordnance, or disposal in place may be used by EOD to eliminate explosive ordnance on ranges.

2-82. EOD may also conduct range clearance operations while deployed to clear ranges of all explosive ordnance in order to safely conduct unit training.

These operations may be conducted on ranges used by several nations with a wide variety of munitions. It is important to include EOD in the planning of unit training while using a foreign nations range.

Remove Stuck Rounds and Download Misfired Munitions

2-83. Removal of stuck rounds in mortars, artillery tubes, and other weapon systems is a routine EOD operation. Prior to requesting EOD assistance for stuck rounds, the using unit must first perform operator and organizational maintenance procedures stated in the weapon's technical manuals.

2-84. EOD may also be used to download misfired munitions from aircraft and armored vehicles weapons systems, both foreign and U.S. If an enemy aircraft or armored vehicle is captured, the capturing unit should immediately request EOD assistance in order to remove all explosive hazards.

2-85. Some EOD procedures require the removal of stuck rounds utilizing explosive techniques that may damage the weapon system. When explosive procedures are used, it is required that the weapon system be taken out of service and tested for stress fractures before returning it to operation.

Remove Ordnance from Battle Damaged Weapon Systems

2-86. EOD inspects battle damaged weapon systems (air and ground, U.S. and foreign) for hazardous conditions. EOD removes explosive ordnance from battle damaged weapon systems and disposes of it. EOD also responds to aircraft accident scenes and clears the scene of explosive ordnance. During response to aircraft accident scenes, special considerations should be made for other hazards not attributed to explosive ordnance such as radiation hazards caused by navigation systems and aircraft components that contain high pressure gasses. Appropriate protective equipment and specialized tools should be utilized.

2-87. If a vehicle or aircraft that contains explosive ordnance catches fire, the area should be evacuated immediately. Ammunition and explosives contained inside or attached to the vehicle or aircraft can detonate. Units should not approach the vehicle or aircraft until EOD has cleared the area of all explosive hazards.

Respond to Chemical and Biological Explosive Ordnance Incidents

2-88. EOD forces respond to U.S. and foreign chemical and biological weapons to identify chemical agents and biological agent precursors, collect samples for further analysis, leak seal and package munitions, render safe, dispose, and conduct emergency personnel decontamination station operations for EOD personnel and their equipment only.

2-89. EOD can identify fillers based on munitions characteristics and nomenclature. EOD does not have the ability to positively identify chemical or biological agents contained within unknown or unidentified non-leaking munitions. CBRNE Response Teams maintain the diagnostic equipment to positively identify chemical fillers within non-leaking munitions. Currently there is no capability to positively identify biological fillers within unidentified non-leaking munitions.

Respond to Depleted Uranium Incidents

2-90. Accidents or incidents involving ammunition or armor containing depleted uranium (DU) pose special challenges. Technical Bulletin 9-1300-278 contains specific guidance for response to DU. EOD Soldiers must be familiar with that guidance before responding to any potential DU incident.

2-91. EOD forces identify DU ordnance and equipment damaged by DU ordnance, detect alpha contamination, control the spread of contamination, package DU ordnance, and conduct emergency contamination control station operations for EOD personnel and their equipment only.

Support Weapons of Mass Destruction Operations

2-92. Weapons of mass destruction are chemical, biological, radiological, or nuclear weapons capable of a high order of destruction or causing mass casualties and exclude the means of transporting or propelling the weapon where such means is a separable and divisible part from the weapon. (JP 1-02) All EOD companies are trained to provide first response to suspected WMD. They act in coordination with more specialized national WMD response assets.

Provide Nuclear Accident and Incident Response and Assistance

2-93. Nuclear accident and incident response and assistance (NAIRA) is intended to minimize loss of life, personal injury, hazardous effects, and destruction of property and to secure classified nuclear material. Procedures described in this section are primarily intended for peacetime use. The DOD may provide support to the lead agency in an accident or incident involving radioactive materials. If required, EOD will respond from the closest military installation.

2-94. EOD forces identify and confirm the presence of nuclear weapons or materials; detect alpha, gamma, and beta hazards; perform actions to prevent a nuclear yield or high explosive detonation; perform initial packaging of materials; conduct emergency contamination control station operations for EOD personnel and their equipment only.

Support to Special Operations Forces

2-95. EOD support to special operations forces requires force packages and personnel requirements which are tailored to fit the specific needs of the supported unit and mission.

Support to special operation forces occurs outside of the Army force generation cycle and does not relieve the requirement for EOD support to ongoing and future operations.

Support to Local, State, and Federal Emergency Response Organizations

2-96. Army EOD is responsible for response to all land mass within CONUS not otherwise designated for coverage by another branch of DOD. Explosive ordnance disposal Soldiers will respond to civil government requests involving UXO, IED, accidents involving military munitions, and other explosive-related incidents in accordance with the Military Munitions Rule (MMR) and locally established Memorandum of Understanding (MOU). EOD can also provide training to civil law enforcement personnel and support to ordnance amnesty programs, (AR 75-14).

Support to Mortuary Affairs Operations

2-97. Immediate recovery and clearance of remains is a priority of all the Services. Mortuary Affairs Soldiers cannot properly recover remains containing UXO and require EOD assistance. Commanders must integrate EOD into their plans for recovery and processing of U.S. and multinational forces remains.

2-98. Every mortuary affairs collection point should include a barricaded, remotely located position to temporarily place remains containing UXO. EOD personnel will X-ray the remains, if necessary, to determine the condition of the UXO. This remote position provides limited protection to ongoing mortuary affairs collection point operations while EOD responds and executes appropriate procedures.

Support to Joint Prisoner of War/Missing in Action Accounting Command

2-99. Joint Prisoner Of War/Missing in Action Accounting Command (JPAC) is a standing organization under the CCDR, Pacific. EOD forces support the JPAC mission of providing the fullest possible accounting of U.S. personnel listed as missing in action worldwide. Most sites that JPAC teams investigate are littered with UXO from military action or aircraft crash sites. EOD forces provide explosive ordnance safety assessment at investigation and recovery sites so JPAC recovery personnel can operate in a safe environment.

Support to Nuclear and Chemical Accident And Incident Control Plans

2-100. A nuclear or chemical accident or incident can cause great damage and have profound political and psychological impact. Control mechanisms, plans, and strict precautions must be maintained to bring the situation quickly under control. Policies and responsibilities for nuclear accident and incident control are outlined in AR 50-5. Policies and responsibilities for chemical warfare accident and incident control are outlined in AR 50-6. More information on CBRN operations can be found in FM 3-11.21.

2-101. The EOD unit must advise the nuclear accident and incident control officer, the chemical accident and incident control officer, or the onsite commander about render safe or disposal procedures and evacuation of the munitions or components. Final disposition of a nuclear weapon or chemical munitions can only be carried out with permission from the nuclear accident and incident control officer, the chemical accident and incident control officer, or the onsite commander.

Integration of Capabilities

2-102. Integrating the special capabilities of EOD organizations requires an understanding of the various capabilities and limitations of the EOD assets available for any given mission. Integration also requires a common understanding of the C2 structure and processes in place to employ the EOD capabilities in unified action. EOD teams do not operate outside of secured areas without appropriate security forces provided by the supported unit. Supported unit planning meetings provide efficient, effective forums for planning operations and coordinating required operational support for EOD companies.

2-103. The commander is able to draw from a force pool of modular EOD units and personnel available to be integrated into joint forces at various echelons. Commanders at all levels must incorporate EOD operations into their planning process.

2-104. EOD force packages must contain adequate capabilities to assure timely and relevant EOD support to the commander. Requirements for EOD increase dramatically during transitions to stability operations and commanders must plan for these changes. Integrating EOD into higher echelon headquarters enables expeditionary action and flexible tailoring of forces to meet changing situations.

Categories

2-105. EOD commanders categorize EOD incidents according to potential threats. Incident categories may later be adjusted depending on the tactical situation, the target, updated intelligence, or field evaluation feedback from the EOD response team onsite. Four incident categories are established in Standardization Agreement (STANAG) 2143, Edition 5.

2-106. Safety is the primary concern when planning EOD operations. Except when faced with a Category A incident, safety of the Soldier is paramount. In all incident categories, the first plan of action should be remote procedures. The four incident categories are described below.

- Category A incidents prohibit unit maneuver and mission capabilities or threaten critical assets vital to the effort and are assigned priority over all other incidents. Category A is assigned to incidents that constitute a grave and immediate threat. Category A incidents are those that can cause mass destruction, widespread contamination, major reduction of combat personnel, or loss of critical facilities or infrastructures. EOD procedures are to be started immediately, regardless of risk to EOD personnel.
- Category B incidents constitute an indirect threat that impair unit maneuver and mission capabilities or threaten critical assets. Items of technical intelligence value are normally assigned to this category. Before EOD operations are started a safe waiting period may be observed to reduce the hazard to EOD personnel. This category correlates with the indirect priority of the incident report.
- Category C incidents constitute a minor threat with potential to reduce unit maneuver and mission capabilities or threaten noncritical assets. EOD personnel normally deal with category C incidents after category A and category B incidents, as the situation permits. Strictly adhere to all wait times and safeties; EOD personnel should expose themselves to the minimum hazards to accomplish these missions.
- Category D incidents constitute no threat and have little or no effect on unit capabilities or assets. Category D ordnance is still lethal and requires EOD response. Mark and leave category D ordnance for EOD assessment.

Priorities

2-107. Priorities give brigade commanders a tool to manage EOD assets in their AO based on the potential threat and needs. These priorities are included within a commander's critical asset list of key assets that have priority of EOD support to maintain essential capabilities and or freedom of maneuver. The EOD company commander synchronizes EOD incident categories with EOD incident priorities for the brigade commander's staff, based on the brigade commander's priorities. Four priorities are established for EOD support.

- Immediate: This priority stops the unit's maneuver and mission capability. It threatens critical assets vital to the mission.
- Indirect: This priority restricts the unit's maneuver and mission capability. It may also threaten critical assets important to the mission.
- Minor: This priority reduces the unit's maneuver and mission capability. It may threaten noncritical assets of value.
- No Threat: This priority has little or no effect on unit capabilities or maneuverability. The item is identified and reported for later action and force protection/public safety awareness.

Explosive Ordnance Disposal 9 Line

2-108. The EOD 9 Line (formally known as the UXO spot report) is the standardized method for the Soldier to report found explosive ordnance. Army battle control system (ABCS) supports this standardized report. The report provides a brief description, location, and recommended priority to assist in determining response categories. The report consists of nine lines, and the reporting unit sends it by the fastest means available.

Line 1	DTG	131200ZAUG04
Line 2	Reporting Unit/EH location	1/7th CAV, EH13221433
Line 3	Contact Method	F400, Sapper 6, CPT Baim
Line 4	Type of Ordnance	82MM mortar 1 ea
Line 5	CBRN contamination	Yes, Soldiers have blisters; M8 paper confirms H
Line 6	Resources threatened	Personnel, mine clearance equipment
Line 7	Impact on mission	Mine clearance operations are stopped
Line 8	Protective measures	Personnel evacuated to 300M; sandbagged barrier constructed
Line 9	Recommended priority	Immediate

Figure 2-3. Example EOD 9 Line.

- Line 1. Date-time group (DTG). The DTG when the item was discovered.
- Line 2. Reporting unit and explosive ordnance location. The unit designation of the reporting unit and the location of the explosive ordnance in an 8-digit grid coordinate.
- Line 3. Contact method. Provide the radio frequency and the call sign, and/or the telephone number and point of contact.
- Line 4. Type of explosive ordnance. Note the size, quantity, type of ordnance (dropped, projected, placed, possible IED, or thrown). Indicate the emplacement method and type of initiation device.
- Line 5. CBRN contamination. If CBRN is present, be as specific as possible. (For example, chemical agent monitor detected G agent at 3 bars; Soldiers are experiencing symptoms of nerve agent; excessive amount of dead wildlife.)
- Line 6. Resources threatened. Report any equipment, facilities, or other assets that are threatened.
- Line 7. Impact on mission. Provide a short description of your current tactical situation and how the presence of the explosive ordnance affects your status (delayed, diverted, cancelled).
- Line 8. Protective measures taken. Describe measures taken to protect personnel and equipment (evacuated to 300M, item marked, sandbag barrier constructed).
- Line 9. Recommended priority (immediate, indirect, minor, or no threat). Ensure that the priority recommended corresponds with the tactical situation described on line 7 of the report (impact on mission). These priorities refer only to the explosive ordnance impact on the current mission. A priority of MINOR or NO THREAT does not mean that the explosive ordnance is not dangerous. (See PRIORITIES from previous section.)

Summary

2-109. The ability of EOD forces to conduct offensive, defensive and stability operations while providing civil support is central to full spectrum operations. Explosive ordnance disposal supports the commanders' mission and the warfighting functions through its four functional areas. Explosive ordnance disposal teams are ready to respond immediately to explosive ordnance incidents in accordance with the commanders' priorities. The EOD Commander coordinates directly with supported units to integrate EOD capabilities into operations.

SECTION 3. COMBAT OPERATIONS

Explosive ordnance disposal personnel play a vital role in support of combat operations. Today, the enemy's weapon of choice is the IED and the demand for EOD is greater than ever. Throughout the wars in Iraq and Afghanistan EOD has had to transform to a highly mobile force able to integrate into all combat formations and operate in any terrain and weather. It is essential that the lessons learned from these conflicts continue to shape the way leaders incorporate EOD into combat operations.

Organization

3-1. Army command and support relationships are the basis for building Army task organizations. Commanders usually establish support relationships in lieu of command relationships between EOD units and non-EOD units, because EOD support is more effective when the commander with the requisite technical and tactical expertise controls the supporting EOD unit. During combat operations, the EOD Company typically operates in direct support of a BCT or BDE, or general support-reinforcing to designated sustainment brigades. The EOD Company provides the BCT or BDE with staff augmentation and planning in order to advise the supported commander on the best implementation of EOD forces.

3-2. Based on mission requirements, the EOD platoon can operate decentralized from the company headquarters and be further task organized to support requirements throughout the BCT AO. Explosive ordnance disposal platoons can provide limited augmentation to battalion HQs with special staff expertise associated with explosive ordnance threats.

3-3. Explosive ordnance disposal platoons typically consist of three EOD response teams. A response team will typically consist of one Team Leader and two team members, but can also operate as two person teams under special circumstances. Response teams are able to operate independently from their platoon HQ, but only for a limited amount of time. For planning purposes an EOD team can respond to six incidents per 24 hour period.

EOD Response

3-4. Explosive ordnance disposal companies provide explosive ordnance response and training to friendly forces in their designated AO. Depending on mission, enemy, terrain and weather, troops and support available, time available, civil considerations (METT-TC) the EOD team may be required to provide response in a variety of different situations. Explosive ordnance disposal team capabilities will change depending on the type of response that is being conducted.

3-5. The EOD Team Leader is the sole operational authority within the inner cordon during an EOD mission. This is to ensure the safety of the EOD team and for all others in the surrounding AO. No personnel or equipment not specifically authorized by the EOD Team Leader will enter or operate within the inner cordon. The EOD Team Leader will ensure the on-scene commander is aware of security requirements, the role of the EOD Team Leader at the incident site, and is notified prior to any controlled detonations during EOD operations. The EOD Team Leader will direct the use and positioning of all Counter Remote Controlled IED Electronic Warfare (CREW), electronic countermeasure (ECM) and communications equipment on site during the conduct of all EOD missions. CREW and communications equipment which are not utilized correctly will result in the reduction or elimination of coverage provided by CREW devices within the inner cordon or within the vicinity of the EOD response. CREW may also interrupt the operation of EOD specific equipment, such as the robot. This operational authority also includes the coordination of the incident site for exploitation and searches as well.

Mounted Response

3-6. Explosive ordnance disposal teams are on standby to respond to EOD 9 Lines in order to support maneuver units operating throughout their AO. Mounted response allows the EOD team to bring all required tools, equipment, and explosives in order to perform RSP and dispose of all explosive ordnance threats in the safest manner possible. Mounted response is the preferred and safest method of response for an EOD team because of the protection of the vehicle and the ability to utilize all protective equipment and tools.

3-7. The Mine Resistant Ambush Protected (MRAP) class of vehicle is central to providing the most protection, while allowing for maneuverability of the EOD team. The MRAP has changed the way EOD teams respond to, and operate, at an incident site. Because of the vehicle's design, teams are able to get closer to the explosive ordnance hazard while staying protected. This allows teams to gain a better understanding of the problem and eliminate the hazard in a more expeditious way.

3-8. Remote operations are the preferred method of performing EOD procedures on an explosive ordnance threat. Explosive ordnance disposal teams may operate with several types of robots depending on the mission. During an EOD incident response, teams will typically operate with a lightweight fast robot. Robots can be deployed from the team's incident control point and be used to perform a reconnaissance of the suspect device, deliver an explosive charge, or carefully render safe the device using the robotic claw. Explosive ordnance disposal robot operators train continuously in order to become highly skilled at performing robotic operations.

3-9. Explosive ordnance disposal technicians are trained to perform X-ray procedures on suspect devices, when possible. This allows the technician to gain an understanding of the design and functioning, and determine the best course of action to defeat the device in the safest manner. Because IEDs are often placed in packages that blend in with their surroundings, it is not uncommon to mistake a harmless item for a possible IED. Correct interpretation of an X-ray can confirm if a suspect device is real or not.

3-10. Explosive ordnance disposal teams utilize a Hook and Line Kit which contains a variety of ropes, pulleys, and other tools that allow an EOD technician to remotely open or move objects while maintaining cover. In order to become proficient with the Hook and Line Kit, an EOD technician must have an understanding of the abilities of the kit and how to properly employ all of the tools.

3-11. Before declaring that a scene is explosively safe, or when the use of a robot is not possible, an EOD technician may have to perform manual procedures on a device. When performing manual procedures, a bomb suit will be worn. Bomb suits provide enhanced protection against blast and fragmentation and have been proven to save lives. A bomb suit will restrict the movements of the EOD technician and the weight of the suit makes it cumbersome to work in. The EOD Team Leader will make the determination if the bomb suit should be worn during a response. Instances where the bomb suit may not be worn include; likelihood of direct enemy fire; device is located in an area where the bomb suit will hinder the movement of the EOD technician; category A incidents.

3-12. Explosive ordnance disposal technicians use a variety of specialized explosive charges that limit collateral damage by focusing the energy of the charge to a specific area of the device, or its container. Specialized charges will often require water and can be very heavy. Explosive ordnance disposal technicians train on placement and delivery techniques in order to get the maximum effectiveness from these charges.

Dismounted Response

3-13. Explosive ordnance disposal teams support dismounted operations during routine patrols or planned missions where explosive ordnance may be encountered. These missions require team members to be especially proficient in warrior tasks and battle drills until their EOD skills are required. Explosive ordnance disposal teams must carry enough food, water, and ammunition for the mission as well as EOD specific equipment and explosives.

3-14. The level of risk for the EOD Soldier is much higher during dismounted operations than mounted operations. The ability to perform remote procedures is limited and the bomb suit is unavailable. There is also less of a variety of specialized EOD equipment and explosive charges. The EOD team must operate in the safest manner possible relying on their training and the experience of the Team Leader.

3-15. When an explosive threat is found, a cordon is set by the patrol around the device. Without the protection of armored vehicles, it is important for Soldiers to utilize the natural terrain or manmade objects as cover until the EOD Team Leader has verified that the threat has been eliminated.

3-16. Dismounted patrols may locate weapons caches that are too large for the EOD team to dispose of, because of the limited amount of explosives that can be carried. This may require resupply of additional demolition materials and reinforcements of additional EOD teams. If a large weapons cache is found, the finding unit will report the location of the cache and await orders.

Route Clearance

3-17. Explosive ordnance disposal teams' support engineer route clearance operations by providing the technical expertise to render safe and dispose of IEDs and UXO that are located. Explosive ordnance disposal teams may incorporate themselves into route clearance patrols or may respond to route clearance patrols after receiving an EOD 9 Line.

3-18. The EOD team can incorporate themselves into the route clearance patrol in several ways. The EOD Team Leader can ride in the Mine Protected Clearance Vehicle (MPCV) in order to advise on the actions to be taken when an explosive threat is encountered. The team members will ride in follow on vehicles with the EOD specific equipment and explosive charges. The Team Leader will keep in contact with the team members and, if necessary, instruct them to perform actions on the explosive threat. In certain instances, the Team Leader may also find it necessary to exit the MPCV in order to perform actions on the explosive threat or collect forensic intelligence after disruption.

3-19. Explosive ordnance disposal teams may also support route clearance patrols by incorporating their response vehicle into the route clearance formation. This allows the EOD team to bring along all equipment and special charges in order to perform actions on a device.

Area Clearance and Minefield Operations

3-20. Area clearance and minefield operations are doctrinally an engineer mission. Explosive ordnance disposal will provide support to these missions in cases where RSPs need to be performed, or the area contains sub-munitions and multiple UXOs. Engineer Explosive Ordnance Clearance Agents (EOCA) may dispose of UXO by detonation if it is found in the EOCA identification guide.

Amnesty and Weapons Buyback Programs

3-21. Explosive ordnance disposal personnel respond to and assist with amnesty collection points to ensure armed or unsafe explosive ordnance is disposed of correctly. Amnesty collection points should be established in a well marked area and at a safe distance from living and working areas. Commanders should utilize EOD personnel when establishing amnesty collection points.

3-22. Weapons buyback programs allow the civilian populace to turn in explosive ordnance and weapons for money. Explosive ordnance disposal personnel support these programs by providing the ability to X-ray, and dispose of any explosive ordnance that is armed, unsafe, or intentionally booby trapped. When establishing a weapons buyback site, commanders must ensure that there is an area established that can be used for emergency destruction of explosive ordnance that is in an unsafe condition, and should ensure that personnel are aware of the possibility of encountering booby trapped munitions.

Weapons Intelligence Teams

3-23. Weapons Intelligence Teams (WIT) are specially trained teams that provide weapons-focused technical intelligence support to operations and are dispatched according to the commander's intelligence collection requirements. WITs focus on the exploitation of IEDs, improvised weapons, other weapons systems, and associated components. WITs may be comprised of EOD personnel, and intelligence and military police personnel that have completed the Weapons Intelligence course.

Post Conflict Considerations

3-24. EOD operations continue after the conflict ends. Explosive ordnance disposal support may include supporting the remediation of the battlefield, and assisting in the retrograde of US munitions by destroying unserviceable or hazardous munitions. Depending on the size of this mission, UXO contractor support may be utilized.

Dependencies

3-25. The EOD Company contains an extremely limited logistic support capability. It is critical that the supported unit understand the requirements of the EOD Company.

The EOD Company must coordinate closely with the brigade, battalion, or Task Force for support. This support includes—

- casualty evacuation,
- vehicle recovery,
- maintenance,

- lodging,
- religious,
- legal,
- force health protection,
- finance and,
- administrative services.

3-26. Explosive ordnance disposal teams do not have the ability to provide security while conducting responses. Explosive ordnance disposal teams rely on security teams to escort them to and from incident sites. Once arriving at the incident site, EOD security will provide protection for the EOD team. Due to mission requirements, EOD technicians may not have the ability to carry their weapons or maintain situational awareness of their surroundings. Explosive ordnance disposal security does not relieve the finding unit of their security responsibilities.

The finding unit must remain on site and maintain a cordon until the EOD Team Leader has declared the area explosively safe. Having a dedicated security team, as opposed to a quick reaction force for EOD, creates a habitual relationship that will allow for a faster response time and greater operational efficiency and effectiveness.

Concept of Communications

3-27. A secure and robust communications system gives the EOD Commander the ability to maintain awareness of current explosive ordnance threats and advise the combatant commander on enemy actions related to explosive ordnance.

A communications system that provides connectivity throughout the AO and across all joint and multinational forces is vital to planning, conducting, and sustaining operations. EOD must maintain communications when its forces are widely dispersed and operating independently from their higher headquarters.

EOD must be able to provide for the timely flow of information in accordance with the commander's priorities with integrated and secure communications capable of line of sight, beyond line of sight, and reach back to DOD and national level agencies to achieve unity of effort. Department of Defense and national level agencies are defined in Appendix D.

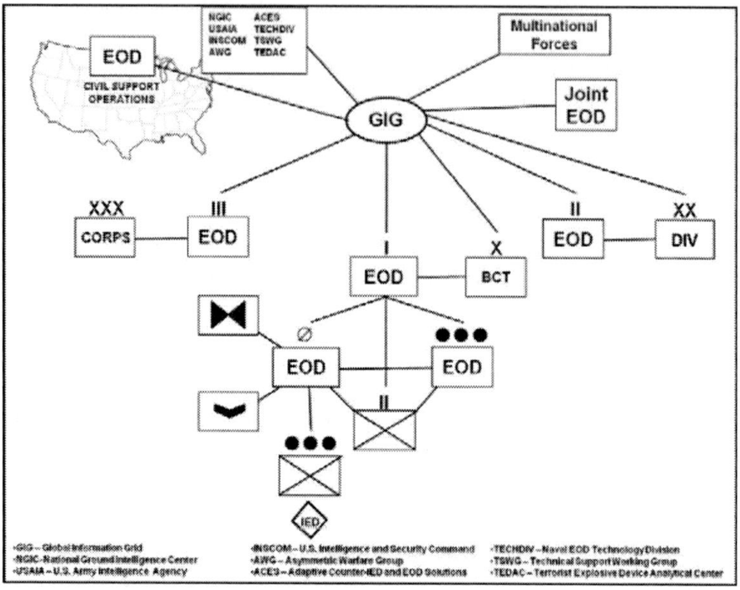

Figure 3-1. EOD Communication Architecture.

Group

3-28. The EOD Group operates at the theater army or corps level; typically collocates the main command post (CP) with the army forces or joint force land component commander joint operations center (JOC) or corps main CP. The EOD Group provides technical direction for information throughout the command and to other selected units. The main CP must be able to communicate with subordinates down to the EOD company level, supported units, and higher headquarters on all required nets as well as joint EOD forces.

3-29. The EOD Group receives EOD incident reports that may include voice, data, digital images, and video from the EOD Battalion. The EOD Group CP organizes the information and delivers it to the JOC, if collocated or using the appropriate information system based on the classification of the information.

3-30. During the movement phase and until the theater matures, the EOD Group CP passes actionable intelligence data to the JOC and selected units using the Defense Information Systems Network services. When the theater matures, the EOD Group CP communicates with the JOC and selected units using the appropriate information system based on the theater G-6 guidance.

3-31. The EOD Group receives requests for EOD support through the local area communications network using the ABCS. Key leaders maintain communication with EOD Group CP, higher headquarters, movement control, and supported units while moving through the AO.

Battalion

3-32. The EOD Battalion operates at the theater army, corps or division level. The EOD Battalion typically collocates its main CP with the corps or division CP. The EOD Battalion provides technical direction for the EOD mission and disseminates EOD technical information throughout the command and other selected units. The main CP must be able to communicate with subordinate units, supported units and higher headquarters on all required networks.

3-33. The battalion receives EOD incident reports that may include voice, data, digital images, and video from the EOD Company. The EOD Battalion organizes the information and delivers it to the EOD Group utilizing the most expeditious means available.

Company

3-34. The EOD Company generally operates at the brigade level. The EOD Company CP may be collocated with the EOD Battalion CP or with the supported unit away from the EOD Battalion. The CP must be able to communicate with subordinate elements down to team level, supported units and higher headquarters on all required networks.

3-35. The EOD Company normally receives EOD incident reports that may include voice, data, digital images, and video from the EOD platoons. The EOD Company organizes the information and delivers it to the EOD Battalion utilizing the most expeditious means available.

Platoon

3-36. The EOD Platoon operates at the maneuver battalion level. The EOD Platoon may be collocated with the EOD Company CP or with a maneuver battalion away from the EOD Company. The platoon CP must be able to communicate with the supported unit as well as the EOD teams.

3-37. The EOD Platoon receives EOD incident reports that may include voice, data, digital images, and video from the EOD teams. The EOD Platoon organizes the information and delivers it to the EOD Company utilizing the most expeditious means available.

Teams

3-38. The EOD team maintains communications with the EOD Platoon as well as the supported unit. The supported unit may task the EOD team with a follow on incident response mission while the team is executing a previously assigned mission. While in route to an incident, the EOD team will communicate with their security element. The team will also communicate with the unit that is at the incident site, so that they can gain a better understanding of the incident site before arrival. Army EOD teams respond to incidents involving all friendly forces within their AO and must be able to establish and maintain communications throughout the response, to include air assets.

3-39. While dismounted, it is essential that the EOD team communicate with each other. The Team Leader will receive information from the on scene commander of the requesting unit and communicate that information to the team and security. This allows the Team Leader to control the scene and direct EOD operations throughout the response.

3-40. The EOD team will create an EOD incident report that may include voice, data, digital images, and

video. The EOD team will deliver it to the EOD Platoon utilizing the most expeditious means available.

EOD Information Systems

3-41. The complexity of current and emerging weapon systems require that EOD have comprehensive reach back capabilities in order to access joint service data repositories and other national level intelligence databases. Additionally, EOD formations at all levels must be able to provide and receive near real-time reporting on weapon systems, explosive ordnance and trends in order to populate databases with mission specific information. The continuous nature of reporting and research, coupled with the distributed footprint of EOD formations necessitates a robust information system. This system must be capable of synchronizing both data and voice communications, as well as plugging into supported unit communications architecture. See Appendix D for more information concerning EOD reach back to national level databases.

Automated EOD Publication System (AEODPS)

3-42. Automated EOD publication system comprises a classified set of electronic publications specific for the EOD mission. The automated EOD publication system also provides tools for planning and decision making within EOD specific tasks and missions. Each publication references an individual piece of ordnance, weapon system, or aircraft, foreign or domestic, and includes general and specific safety precautions, positive identification criteria, render-safe, and disposal procedures. Additional publications explain the use of EOD specific tools and TTPs. Explosive ordnance disposal publications are published on the automated EOD publication system and are distributed on DVDs to all EOD specific units within the joint service community. The overall classification level of the material contained with the automated EOD publication system is SECRET, with appropriate caveats for specific topics and information.

3-43. Access to automated EOD publication system is limited to those individuals with a need-to-know. EOD units are not authorized to release this information to non-EOD personnel. Requests for automated EOD publication system must be routed through the release authority, Naval EOD Technology Division.

Summary

3-44. Explosive ordnance disposal's application of decentralized control effectively integrates a high level of technical expertise, a streamlined command and force structure, as well as timely and responsive support. The senior EOD organization commander must be prepared to modify the duties and responsibilities of the staff depending upon which organization the EOD unit is supporting. Decisions regarding establishing joint EOD operations are made by the JTF or combatant commander based on mission analysis, EOD forces available, and requirements. The main CP must be able to communicate with subordinates, supported units, and higher headquarters on all required nets. Additionally, EOD formations at all levels must be able to provide and receive near real-time reporting on weapons systems, explosive ordnance and trends in order to utilize databases with mission specific information.

APPENDIX A. FORCE STRUCTURE

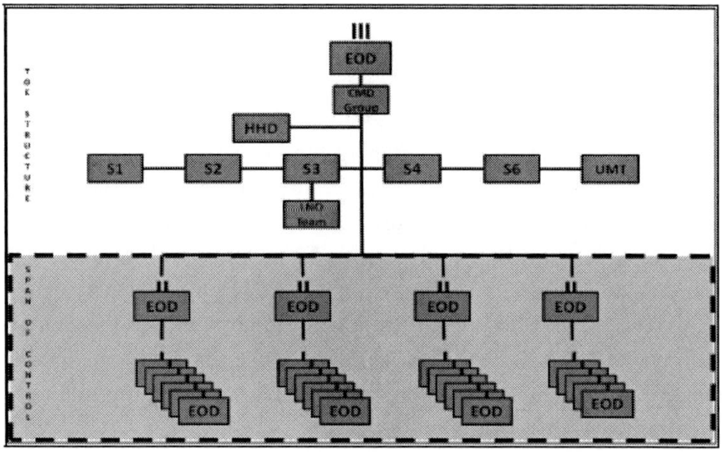

Figure A-1. EOD Group Structure.

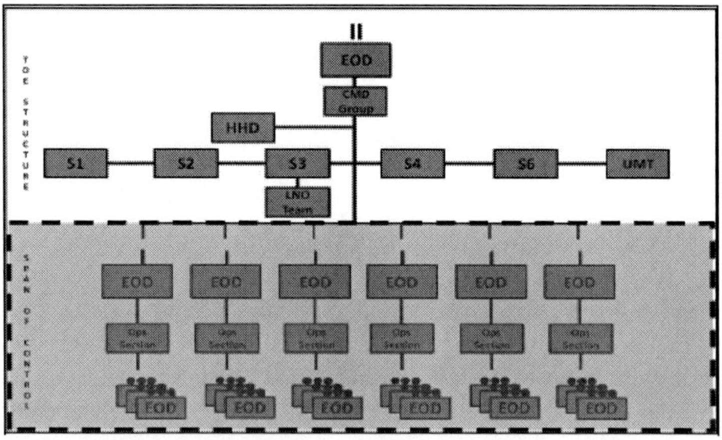

Figure A-2. EOD Battalion Structure.

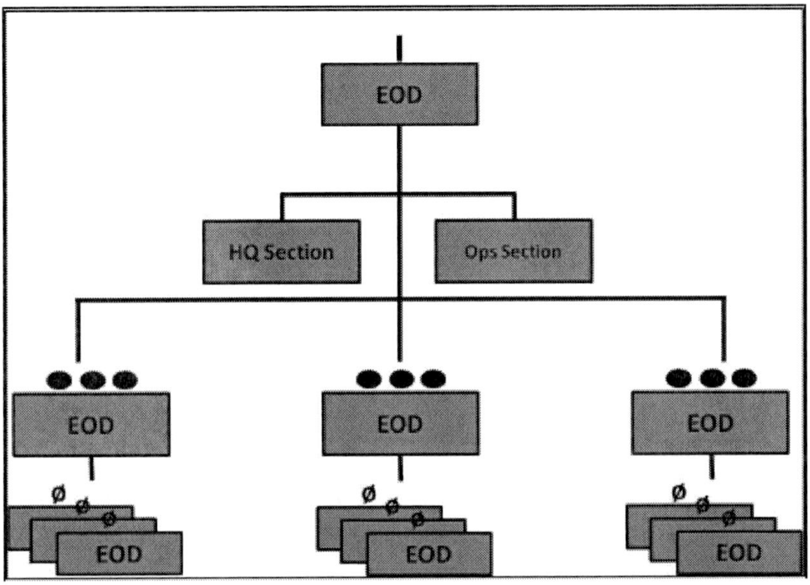

Figure A-3. EOD Company Structure.

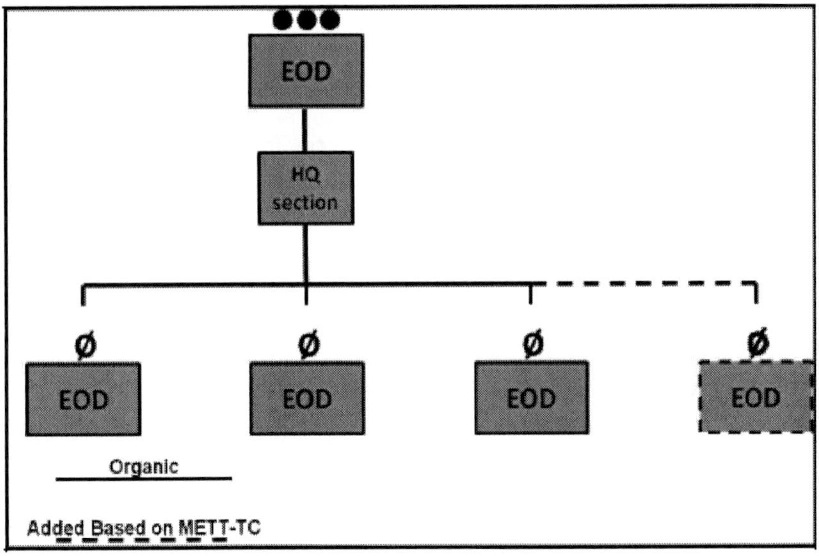

Figure A-4. EOD Platoon Structure.

APPENDIX B. UNITED STATES OPERATIONAL GUIDELINES

This appendix provides excerpts from the Military Munitions Rule (MMR) (Title 40 Code of Federal Regulations [CFR] part 260, et seq.) as amended in 1997 and Resource Conservation and Recovery Act (RCRA) of 1980 only as a guide. The MMR and RCRA must be used when developing local policies and procedures. This guidance does not supersede any local or state requirements that may apply to any installation. The EOD unit should coordinate with state and local agencies through the installation environmental office. A MOU should be prepared for all local and state agencies.

Homeland Operational Guidelines

B-1.DOD provides homeland defense EOD support to civil authorities within the United States. Each military service is responsible for providing EOD support to those DOD installations for which they are responsible and for those military munitions in their possession. In emergencies, the closest EOD unit should respond immediately with the understanding the responsible service retains control. EOD units will maintain a 24-hour phone watch to receive civil authority reports of a known or suspected discovery of munitions and explosives of concern in the public domain and other requests for EOD support.

B-2.EOD units do not respond to requests for support directly from private citizens or organizations. Private citizens and organizations must direct their requests to an authorized official (local law enforcement) who will in-turn report the incident and request assistance from the supporting EOD unit.

B-3.The DOD military service that first becomes aware of an incident involving either military munitions or munitions or explosives of another federal agency will take immediate action to prevent or limit damage or injury. Such actions include gathering information about the situation from local authorities, environmental regulators and safety officials, property owners and others, as appropriate.

B-4.The Department of Defense Explosive Safety Board (DDESB) is responsible for establishing explosives safety policy under authority delegated to it by the Secretary of Defense.

B-5.Environmental regulators and safety officials may have independent authority, responsibility for, and the technical expertise to evaluate public safety and environmental aspects of response actions, to include explosives or munitions emergency responses. Typically, EOD and environmental regulators and safety officials work closely to manage any potential adverse impacts on the scene. Situations that would require environmental regulators or safety officials to raise concerns about the potential impact of EOD's intended course of action (COA) to higher levels of DOD authority should be the exception, not the rule.

B-6. If local on-scene civil authorities (safety officials, law enforcement, environmental regulators) have serious concerns about the potential impact of an intended COA they should first discuss the concern with the Team Leader. If no resolution can be found then elevate their concern to a higher level of authority. Army EOD personnel and local authorities work closely to manage and minimize potential adverse impacts. However, if a change in response action from local authorities would conflict with EOD explosives safety requirements or procedures, the EOD Team Leader shall implement protective measures necessary to ensure the site remains safe and secure until the conflict is resolved. EOD personnel are not allowed to take any action inconsistent with approved EOD procedures or that would place them at added risk.

Munitions Rule, EPA, And OSHA Considerations

B-7. Two difficult challenges EOD units face are to protect the environment while performing the EOD mission and to follow U.S. and host nation environmental rules and regulations.

B-8.Many Federal, state, local, and HN laws now hold commanders legally responsible for environmental damage caused by inadequate planning, supervision of operations, or training. This could mean fines and/or imprisonment if convicted of environmental violations.

B-9. To avoid adverse environmental impact when planning or executing operations, refer to TC 3-34.489 and FM 3-34.5. Also, be familiar with provisions and requirements of AR 200-1.

B-10. It is vital that EOD Soldiers understand the Federal and state environmental protection laws of the area in which they are working. The individual Soldier may be held responsible for actions under these laws if the Soldier is negligent in the performance of duties and fails to follow EPA guidelines.

Environmental Protection and Compliance Policy

B-11. The EPA has finalized regulations that clarify when conventional and chemical military munitions become a hazardous waste under the RCRA. Additionally, this rule amends existing regulations regarding emergency responses and RCRA manifest requirements.

B-12. Army policy and procedures for complying with the RCRA (Title 42 USC section 3251) are found in AR 200-1. Each department of the Federal Government engaged in any activity resulting in the disposal or management of solid or hazardous waste shall be subject to and comply with all federal, state, interstate, and local requirements, both substantive and procedural (including any requirements for permits and reporting). Section 6001 of the RCRA.

B-13. AR 200-1 explains Federal requirements of the RCRA and the Occupational Safety and Health Administration (OSHA) Act (Title 29 CFR part 1910.120 and Title 29 CFR part 1200) as they apply to the management, handling, transportation, storage, and disposal of conventional explosive ordnance. AR 200-1, paragraphs 5-1 and 5-2a requires all Army activities to comply with legally applicable and appropriate Federal, state, and local regulations, both substantive and procedural requirements, for managing, generating, treating, storing, disposing, and transporting hazardous and solid waste. This includes the terms and conditions of state and Federal hazardous and solid waste permits.

Military Munitions Rule

B-14. DOD has adopted the Military Munitions Rule (Title 40 CFR Part 260, et seq.) which defines when conventional and chemical military munitions become solid wastes potentially subject to hazardous waste regulations; and establishes procedures and management standards for waste military munitions. Although the Military Munitions Rule defines explosives or munitions emergency response specialists broadly, within DOD, only EOD personnel will respond to military or civilian authority requests for support of an explosives or munitions emergency.

B-15. Emergency Response Recordkeeping. The MMR establishes a new RCRA record-keeping requirement for all explosives or munitions emergency responses. The responding unit must keep these records for at least 3 years. The content, which is similar to existing EOD records, includes:

- Date and time of response.
- Names and grade of individuals who respond.
- Type and description of the munitions (waste) addressed.
- Disposition of the munitions.
- A written copy of the emergency permit, if issued, and all related documents

B-16. Explosives or munitions emergencies that involve or potentially involve chemical warfare materials present specific and complex challenges. Such challenges may include compliance with the Chemical Warfare Convention and Chapter 32 of title 50 United States Code. During such responses, safety, particularly with regard to the potential hazards associated with chemical agents, is the primary consideration in response planning. Whenever circumstances allow, without compromising safety, coordination with federal, state, and civilian authorities, local emergency responders (e.g., ambulance, and other medical support), and community outreach will be priorities.

B-17. Once EOD determines that an explosives or munitions emergency involve munitions with an unknown liquid filler, chemical munitions, or other recovered chemical warfare material (RCWM), the need for support from and close coordination with other DOD agencies on the response action is mandatory. This is particularly true for the final disposition of RCWM. Such agencies include, but are not limited to the Program Manager for Non-Stockpile Chemical Material, and the Army Secretariat and OSD. In addition, the need for coordination and collaboration with state and federal authorities, to include environmental regulators and safety officials, on the response is necessary.

B-18. EOD's immediate responsibility is to ensure the protection of human health and the environment and that the RCWM or recovered munitions with unknown liquid filler is properly over packed and secured, either at the location of discovery or at the closest military installation. Once these protective measures are in place and the supported activity or EOD makes required notifications, DOD will take those steps necessary to plan for required assessments and disposition.

B-19. The Army further characterizes explosives or munitions emergencies as either level 1 or level 2 emergency responses. These emergency responses can occur anywhere, to include off-post locations. EOD will assume that requests from authorized officials to respond to a discovery of military munitions will require an immediate response. Accordingly, dispatched EOD teams respond to all such requests as if a Level 1 emergency exists.

The responding EOD Team will, upon arrival at the incident site make contact with the requesting local civil agency and ensure that a signed MOU exists between the US Army and the requesting authorities. If one does not exist then the EOD Team Leader will ensure that one is completed and signed before actions are taken. If there is an immediate threat to life the EOD Team Leader may take action before the confirmation of a signed MOU.

- Level 1–Immediate Response. Level 1 responses are immediate responses required to address an actual or potential imminent threat to human health (including safety) or the environment (including property), as determined by the lead on-site explosives or munitions emergency response specialist. Level 1 responses are exempt from federal hazardous waste management generator, transporter, and permitting requirements and may involve locations both on and off DOD installations. EOD personnel will use established procedures and good judgment to determine whether a situation requires a Level 1 response. Only lead on-site DOD explosives or munitions emergency response specialists (that is, EOD emergency response specialist) may determine and terminate Level 1 responses.
- Level 2– Imminent and Substantial Endangerment Response. Level 2 response is a planned response by installation or government. It addresses actual or potential threats posed by explosives or munitions to people or property, where an immediate response to an imminent and substantial endangerment is not required. An emergency permit may be required before conducting Level 2 response actions; the lead on-site EOD emergency response specialist should ensure consultation with the applicable installation's environmental office or the Regional Environmental Coordinator office to determine any reporting or permitting requirements. Level 2 responses may involve locations both on and off DOD installations.

B-20. Level 1 emergency response actions are exempt from RCRA, permitting Department of Transportation (DOT) manifesting for movement to a safe disposal area, and other substantive requirements under Title 40 CFR part 264.1(g)(8), part 265.1(c)(11) and part 270.1(c)(3). However, level 2 emergency response actions may be subject to emergency permitting and other requirements. During an emergency response, the appropriate Army activity (e.g., installation or district commander, environmental office) will consult with environmental regulators and safety officials on the need for permits and fulfilling other requirements.

Resource Conservation And Recovery Act Compliance (RCRA)

B-21. A section in Title 40 CFR 270.1 (c) states that a RCRA permit is required for owners and operators of any facility for the "treatment, storage, and disposal" of any solid or hazardous waste. The definitions of solid and hazardous waste are outlined in 40 CFR 270.2.

B-22. AR 200-1, requires the installation commander to enter into compliance agreements and consent orders to achieve RCRA compliance. Army installations will obtain a RCRA permit to treat, store, and dispose of solid and hazardous waste. The installation commander will be considered the hazardous waste generator, as defined in 40 CFR 260.10, for the purposes of signing RCRA reports and records and for getting an EPA identification number, as outlined in 40 CFR 262.12, paragraph 1-25 f (7).

B-23. All personnel involved in solid or hazardous waste operations or activities must be trained in accordance with 40 CFR 264.16 and 29 CFR 1910.120. This training must be documented and updated at least annually, unless otherwise stated as part of the RCRA permit. Currently, in accordance with the guidance from DA Office for Environment, Safety, and Occupational Health, completion of EOD school meets the requirements of 40 CFR 264.16 and 29 CFR 1910.120.

B-24. Under the RCRA, a generator of a hazardous waste is responsible for that waste from its "cradle to its grave." As defined in 40 CFR 260.10, a generator of a hazardous waste is "any person whose act or process produces hazardous waste identified or listed in 40 CFR 261 or whose act first causes a hazardous waste to become subject to regulation." Conventional explosive ordnance will be considered hazardous waste under the following conditions:

- An authorized official records a written determination that the conventional explosive ordnance will be discarded.
- Custodians of the conventional explosive ordnance receive the written determination to discard the conventional explosive ordnance and, therefore, are subject to RCRA regulations.

B-25. The authorized official when making a determination to discard must consider applicable facts and circumstances for each situation. The following guidelines should be used when making this determination:

- A determination to discard excess conventional explosive material that is safe and stable in normal logistical environments by military standards may be made only after all efforts have been exhausted to reuse, recycle, recover, or sell such material.
- A determination to discard conventional explosive ordnance that may be unsafe or unstable to store or transport should be made by an authorized official after conducting appropriate testing or inspection, if conditions allow, or if it is readily apparent that there is no reasonable alternative to discarding the material.

B-26. Generally, conventional explosive ordnance manufacture, assembly, testing, training, intended use, or range management does NOT constitute hazardous waste as regulated by the RCRA. However, some wastes generated by these operations may be subject to RCRA regulations.

B-27. According to this guidance, conventional explosive ordnance that is safe for transportation and has not been classified as a hazardous waste can be transported, according to DOT and DA guidelines.

B-28. In accordance with RCRA regulation 40 CFR 261.3 (c) (1), once a waste has been identified as a solid or hazardous waste, it will remain a hazardous waste until—

- The waste no longer exhibits a hazardous waste characteristic as described in 40 CFR 261, subpart C.
- The waste has been specifically excluded by regulation (such as being unlisted).

B-29. Many Federal and State/Local environmental laws now hold commanders legally responsible for environmental damage caused by inadequate planning, supervision of operations, or training.

To avoid adverse environmental impact when planning or executing operations, refer to TC 3-34.489, the Commander's Guide to Environmental Management.

Also, be familiar with provisions and requirements of AR 200-1.

Storage, Treatment, and Disposal of Nondefense Toxic and Hazardous Materials

B-30. Except as otherwise provided in this section, the Secretary of Defense may not permit the use of an installation of the Department of Defense for the storage, treatment, or disposal of any material that is a toxic or hazardous material and that is not owned either by the Department of Defense or by a member of the armed forces (or a dependent of the member) assigned to or provided military housing on the installation.

B-31. The Secretary of Defense shall define by regulation what materials are hazardous or toxic materials for the purposes of this section, including specification of the quantity of a material that serves to make it hazardous or toxic for the purposes of this section. The definition shall include materials referred to in section 101(14) of the Comprehensive Environmental Response, Compensation, and Liability Act of 1980 (42 U.S.C. 9601(14)) and materials designated under section 102 of that Act (42 U.S.C. 9602) and shall include materials that are of an explosive, flammable, or pyrotechnic nature. (b) Subsection (a) does not apply to the following:

- The storage, treatment, or disposal of materials that will be or have been used in connection with an activity of the Department of Defense or in connection with a service to be performed on an installation of the Department for the benefit of the Department.
- The storage of strategic and critical materials in the National Defense Stockpile under an agreement for such storage with the Administrator of General Services.
- The temporary storage or disposal of explosives in order to protect the public or to assist agencies responsible for Federal, State, or local law enforcement in storing or disposing of explosives when no alternative solution is available, if such storage or disposal is made in accordance with an agreement between the Secretary of Defense and the head of the Federal, State, or local agency concerned.
- The temporary storage or disposal of explosives in order to provide emergency lifesaving assistance to civil authorities.
- The disposal of excess explosives produced under a Department of Defense contract, if the head of the military department concerned determines that, in each case, an alternative feasible means of disposal is not available to the contractor, taking into consideration public safety, available resources of the contractor, and national defense production

requirements.

- The temporary storage of nuclear materials or nonnuclear classified materials in accordance with an agreement with the Secretary of Energy.
- The storage of materials that constitute military resources intended to be used during peacetime civil emergencies in accordance with applicable Department of Defense regulations.
- The temporary storage of materials of other Federal agencies in order to provide assistance and refuge for commercial carriers of such material during a transportation emergency.
- The storage of any material that is not owned by the Department of Defense if the Secretary of the military department concerned determines that the material is required or generated in connection with the authorized and compatible use of a facility of the Department of Defense, including the use of such a facility for testing materiel or training personnel.
- The treatment and disposal of any material that is not owned by the Department of Defense if the Secretary of the military department concerned determines that the material is required or generated in connection with the authorized and compatible use of a facility of that military department and the Secretary enters into a contract or agreement with the prospective user that—
    - is consistent with the best interest of national defense and environmental security; and
    - provides for the prospective user's continued financial and environmental responsibility and liability with regard to the material.
- The storage of any material that is not owned by the Department of Defense if the Secretary of the military department concerned determines that the material is required or generated in connection with the use of a space launch facility located on an installation of the Department of Defense or on other land controlled by the United States.

B-32. The Secretary of Defense may grant exceptions to subsection (a) when essential to protect the health and safety of the public from imminent danger if the Secretary otherwise determines the exception is essential and if the storage or disposal authorized does not compete with private enterprise.

B-33. The Secretary may assess a charge for any storage or disposal provided under this section. Any such charge shall be on a reimbursable cost basis.

B-34. In the case of storage under this section authorized because of an imminent danger, the storage provided shall be temporary and shall cease once the imminent danger no longer exists. In all other cases of storage or disposal authorized under this section, the storage or disposal authorized shall be terminated as determined by the Secretary.

APPENDIX C. SUPPORT TO THE US SECRET SERVICE

Guidelines

C-1. As directed by AR 75-15, EOD personnel are provided to the US Secret Service (USSS), the Department of State (DOS), and the Department of Justice (DOJ) upon request. The procedures in this appendix address USSS missions because they are most frequent. However, procedures are generally the same for the other organizations.

C-2. The primary responsibility of the EOD personnel supporting the USSS is to detect hazards to the principal (the VIP or dignitary) during a visit, and advise the USSS so that the principal may be evacuated. The primary responsibility for a device, if one is found, lies with the law enforcement agency with jurisdiction for the area or the offense.

Mission Receipt

C-3. VIP support-mission assignments are passed to EOD companies by the EOD Group or Battalion. Upon receipt of the assignment, the senior EOD Soldier will contact the special agent in charge of the missions for any special information or instructions. This will include the required arrival time for the team, where to meet the special agent in charge or the Technical Security Division (TSD) representative, and what special equipment and clothing are needed.

Pre-Mission Preparation

C-4. Prior to the receipt of any VIP-support mission, the EOD company commander will ensure the teams have the items required to perform the mission. They need the Hazardous Device Countermeasures Manual (HDCM), published and furnished by the USSS, and EOD credentials for personnel. They also need identification (lapel) pins in the current series, official vehicle placards, official passports, and a tool kit. The senior EOD Soldier and team members will prepare for a mission by reviewing AR 75-15, the HDCM, as well as local policies.

Mission Execution

C-5. Upon arrival at the designated location, the senior EOD Soldier reports to the USSS representative. The USSS should provide the tentative mission schedule, the areas to be searched, and the route and itinerary to be followed by the dignitary, including alternate routes. Information should also be provided on the location and room numbers of the security room, the package holding area, and the communications room; the EOD capabilities of local authorities; and the threat summary.

C-6. Before and during the dignitary's visit, the senior EOD Soldier surveys all areas to be covered and plans for the search. The senior EOD Soldier recommends any requirements needed to support the EOD teams.

C-7. Normally, the responsible USSS agent plans the search and designates the explosives holding area and evacuation routes. This agent is a communications link for the EOD team. However, in the absence of a TSD agent or other designated representative, the senior EOD Soldier may have to do those duties. The responsible agent relies on the senior EOD Soldier for technical advice and recommendations about explosives. Therefore, the senior EOD Soldier must carefully consider recommendations before giving them to the responsible agent. The recommendations will be used in the coordination and direction of the actual search.

C-8. Before the search, the senior EOD Soldier meets with the USSS agent and local law enforcement officials. At the meeting, they determine the available disposal facilities and local bomb squad capabilities and responsibilities. They also determine the evacuation routes between the visit location and disposal facilities. When established disposal facilities are not available, they select a temporary site and select routes to the disposal sites that avoid crowds and main traffic areas. They also arrange for an escort vehicle in case it is necessary to move an explosive device. Explosives holding areas must as far away as practical from the areas the VIP is scheduled to visit.

C-9. Search procedures for facilities and adjacent areas are outlined in the HDCM. Security of search and post search areas is the responsibility of the USSS. EOD personnel will not be used to secure areas.

C-10. EOD personnel advise the USSS agent when the bomb search is completed. The advance agent then determines where EOD personnel are to wait. The senior EOD Soldier furnishes technical advice for identifying an explosive or incendiary device. The senior EOD Soldier also notifies the agent of the findings. If necessary, this agent helps carry out the evacuation plan.

C-11. When the principal arrives, EOD personnel may, at the discretion of the responsible USSS agent, be assigned to that agent during the visit to provide immediate on-the-scene technical advice. If an explosive item is found, EOD personnel may be charged with helping the responsible agent start evacuation procedures. During the actual visit, any extra EOD personnel may be put on standby status.

C-12. If a bomb, incendiary device, or any suspected item is found during a search, it must be reported immediately to the USSS. Even though speed is critical in reporting a situation of this type, care and discretion must be used in reporting such an incident. Under no circumstances should the find be reported to anyone other than a representative of the USSS.

C-13.During the performance of the mission, every effort must be made to divert public attention from EOD activities. EOD should immediately refer any inquiries about EOD support activities to the responsible USSS representative.

Post Mission Actions

C-14. When the mission is completed, the senior EOD Soldier has an exit briefing with the agent in charge. The purpose is to discuss any problems that came up during the mission.

Any unresolved problems are immediately reported to the EOD company commander.

Just before the EOD team departs, the senior EOD Soldier contacts one of the following to determine if any missions are pending: the USSS coordinating center (when in operation), the EOD Company, or the senior EOD Soldiers home unit.

Appendix D. Supporting Organizations

The complexity of current and emerging weapons systems require that EOD have comprehensive reach back capabilities up to the DOD and national levels. Additionally, EOD teams must have reach across and feedback capabilities to provide and receive near real time, relevant reporting on weapon systems, explosive ordnance and trends in order to populate databases with the most current data.

National Ground Intelligence Center (NGIC)

D-1. NGIC produces and disseminates all-source integrated intelligence on foreign ground forces, systems, and supporting combat technologies to ensure that U.S. Forces have a decisive edge on any battlefield. NGIC supports U.S. Army Forces during training, operational planning, deployment, and redeployment. NGIC maintains a counter IED targeting program CITP portal on the SECRET Internet Protocol Router Network Web site that provides information concerning IED activities and incidents and NGIC IED assessments. In the IED fight, NGIC increases the capability of the coalition force to collect technical intelligence and provide dedicated intelligence fusion to support counter-insurgency operations.

United States Army Intelligence Agency (USAIA)

D-2. USAIA is a field-operating agency of the Defense Communication System DCS that produces and disseminates intelligence information on foreign ground forces and their weapon systems. USAIA also provides threat analysis and related projections to the combat development community of the Army.

United States Army Intelligence and Security Command (INSCOM)

D-3. INSCOM conducts a wide range of production activities, ranging from intelligence preparation of the battlefield IPB to situation development, including signal intelligence analysis, imagery exploitation, and science and technology intelligence production.

Asymmetric Warfare Group (AWG)

D-4. The AWG conducts operations in support of joint and Army Forces commanders to mitigate and defeat specified asymmetric threats. The AWG assists in exploitation and analysis of asymmetric threats and provides advisory training for in-theater or predeployment forces.

G-38, Adaptive Counter-Improvised Explosive Device and Explosive Ordnance Disposal Solutions (ACES) Division

D-5. The G-38 Division provides staff planning for assigned EOD strategic requirements in support of the geographic combatant commander's operational plan. The G-38 Division accomplishes the planning by using the Joint Operation Planning and Execution System (JOPES) and coordinates the planning effort with the combatant commander's EOD staff officer.

D-6. G-38, within HQDA G3/5/7, manages the Army EOD Program in order to rapidly man, train, equip, and organize Army formations with the inherent ability to apply and defeat emergent asymmetric threats and adaptive networks including Weapons of Strategic of Influence and asymmetric weapons, such as the IED, in support of full spectrum operations.

Naval Explosive Ordnance Disposal Technology Division (NAVEODTECHDIV)

D-7. NAVEODTECHDIV exploits technology and intelligence to develop and deliver EOD information, tools, equipment, and life cycle support to meet the needs of joint service EOD operating forces and other specified customers. NAVEODTECHDIV manages the EOD database and provides a 24/7 reach back capability for UXO/IED/CBRN topics.

Technical Support Working Group (TSWG)

D-8. The TSWG is the U.S. national forum that identifies, prioritizes, and coordinates interagency and international research and development requirements for combating terrorism. The TSWG rapidly develops technologies and equipment to meet high priority needs of the combating terrorism community and addresses joint international operational requirements through cooperative research and development with major allies.

Terrorist Explosive Device Analytical Center (TEDAC)

D-9.The mission of TEDAC is to prevent potential IED attacks by coordinating and managing the unified efforts of law enforcement, intelligence, and military assets to technically and forensically exploit all terrorist IEDs worldwide of interest to the U.S. government. The information and intelligence derived from the exploitation of terrorist IEDs is used to provide actionable intelligence to anti-terror missions and to help protect the U.S. military and coalition assets around the globe.

References

Army Publications

ADP 3-0, *Unified Land Operations,* 10 October 2011.
AR 5-9, *Area Support Responsibilities,* 16 October 1998.
AR 50.5, *Nuclear Surety,* 1 August 2000.
AR 50.6, *Chemical Surety*, 28 July 2008.
AR 75-14, *Interservice Responsibilities for Explosive Ordnance Disposal,* 14 February 1992.
AR 75-15, *Policy for Explosive Ordnance Disposal,* 22 February 2005.
AR 200-1, *Environmental Protection and Enhancement,* 13 December 2007.
AR 525-27, *Army Emergency Management Program*, 13 March 2009.
ATTP 3-90.15, *Site Exploitation Operations,* 8 July 2010.
ATTP 4-21.16, *EOD Multi-Service Tactics, Techniques, and Procedures for Explosive Ordnance Disposal,* 20 September 2011.

ATTP 4-32.2, *UXO Multi-Service Tactics, Techniques, and Procedures for Unexploded Ordnance,* 20 September 2011.

FM 3-11.21, *MTTP for Chemical, Biological, Radiological, and Nuclear Consequence Management Operations,* 1 April 2008.

FM 3-34.5, *Environmental Considerations,* 16 February 2010. FM 3-90, Tactics, 4 July 2001.

TB 9-1300-278, *Guidelines for Safe Response to Handling, Storage, and Transportation Accidents Involving Army Tank Munitions or Armor which Contain Depleted Uranium,* 21 July 1996.

TC 3-34.489, *The Soldier and the Environment*, 8 May 2001.

Department of the Army Forms

DA forms are available on the APD web site (www.apd.army.mil).

DA Form 2028, *Recommended Changes to Publications and Blank Forms.*

Joint and Allied Publications

CJCSI 3207.1B, Military Support to Humanitarian Mine Actions, 1 November 2008. JP 1-02, DOD Dictionary of Military and Associated Terms, 8 November 2010. JP 3-0, Joint Operations, 11 August 2011.

STANAG 2143, ED. 5, Explosive Ordnance Reconnaissance/Explosive Ordnance Disposal (EOR/EOD), 16 September 2005.

Department of Defense Publications

DOD 3150.8-M, *Nuclear Weapon Accident Response Procedures (NARP),* 22 February 2005.

DOD Directive 3025.13, *Employment of DOD Capabilities in Support of the U.S. Secret Service (USSS), Department of Homeland Security (DHS),* 8 October 2010.

DOD Directive 3025.18, *Defense Support of Civil Authorities (DSCA)*, 29 December 2010. DOD Instruction 6055.17, *DOD Installation Emergency Management (IEM) Program,* 13 January 2009.

Codes of Federal Regulations

Title 10, United States Code, *Armed Forces.*
Title 18, United States Code, *Crimes and Criminal Procedure.*
Title 29, United States Code, *Labor.*
Title 32, United States Code, *National Guard .*
Title 40, United States Code, *Public Buildings, Property, and Works.*
Title 42, United States Code, *The Public Health and Welfare.*
Title 50, United States Code, *War and National Defense.*

Tables of Organization and Equipment

TOEs are available online at: https://fmsweb.army.mil/unprotected/splash/.

TOE 09446R000, *Headquarters, Headquarters Detachment, Ordnance Battalion (EOD),* 5 March 2010.

TOE 09447A000, *Ordnance Company Explosive Ordnance Disposal (EOD),* 21 June 2010. TOE 09447GB00, *Platoon Ordnance Company (EOD),* 2 February 2007.

TOE 09627R000, *Headquarters, Headquarters Detachment, Ordnance Group, Explosive Ordnance Disposal (EOD),* 19 February 2010.

TOE 09747L000, *Ordnance, Explosive Ordnance Disposal (EOD) Company (CONUS Support),* 22 June 2010.

TOE 09747R000, *Explosive Ordnance Disposal, Weapons of Mass Destruction (WMD) Company,* 21 September 2009.

Helpful Websites

Army Publishing Directorate http://www.apd.army.mil/.
Code of Federal Regulations http://www.gpoaccess.gov/cfr/.
Department of Defense http://www.defense.gov/.
Department of Homeland Security http://www.dhs.gov/index.shtm.
Joint Electronic Library http://www.dtic.mil/doctrine/.

NATO Standardization Agreements http://www.nato.int/cps/en/natolive/stanag.htm.

Naval EOD Technology Division http://www.navsea.navy.mil/nswc/eodtechdiv/default.aspx.

United States Environment Protection Agency

http://www.epa.gov/epawaste/inforesources/online/index.htm.

U.S. Army Ordnance Corps, EOD Directorate http://www.goordnance.army.mil/.

Other Pubications

Homeland Security Presidential Directive-12, *Policies for a Common Identification Standard for Federal Employees and Contractors,* 27 August 2004.

Resource Conservation and Recovery Act of 1980 http://epw.senate.gov/rcra.pdf.

GLOSSARY

Acronyms and Abbreviations

ABCS	Army battle control system
AEODPS	Automated EOD Publication System
AO	area of operations
AR	Army Regulation
ASP	ammunition supply points
ATTP	Army Tactics Techniques and Procedures
BCT	brigade combat team
C2	command and control
CBRN	chemical, biological, radiological, and nuclear
CCDR	Combatant Commander
CEA	captured enemy ammunition
CFR	Code of Federal Regulations
CJTF	combined joint task force
COA	course of action
CONPLAN	concept plan
CONUS	Continental United States
CP	Command post
CREW	Counter Remote Controlled IED Electronic Warfare
DDESB	Department of Defense Explosive Safety Board
DHS	Department of Homeland Security
DOD	Department of Defense

DOJ	Department of Justice
DOS	Department of State
DOT	Department of Transportation
DSCA	Defense Support of Civil Authorities
DTG	Date-time group
DU	depleted uranium
ECM	electronic countermeasure
EOCA	Explosive Ordnance Clearance Agent
EOD	Explosive Ordnance Disposal
FM	Field Manual
GIG	Global Information Grid
HDCM	Hazardous Device Countermeasures Manual
HMA	Humanitarian Mine Action
IED	Improvised Explosive Device
IEM	Installation Emergency Management
ISR	Intelligence, surveillance, reconnaissance
JEODVIPPSA	Joint EODVIP Protection Support Activity
JFC	Joint force commander
JOC	joint operations center
JP	Joint Publication
JPAC	Joint Prisoner Of War/Missing In Action Accounting Command
JTF	Joint task force
MEB	maneuver enhancement brigade
METT-TC	mission, enemy, terrain and weather, troops and support available, time available, civil considerations
MMR	Military Munitions Rule
MOU	Memorandum of Understanding
MPCV	Mine Protected Clearance Vehicle
MRAP	Mine Resistant Ambush Protected
MTTP	Multi-Service Tactics, Techniques, and Procedures
NAIRA	Nuclear accident and incident response and assistance
NARP	Nuclear Weapon Accident Response Procedures
NATO	North Atlantic Treaty Organization
NCO	noncommissioned officer
OE	operational environment
OPCON	operational control
OPLAN	operation plan
OPORD	operation order

OSHA Occupational Safety and Health Administration
RCRA Resource Conservation and Recovery Act
RCWM recovered chemical warfare material
RSP render safe procedures
SOP standard operating procedures
STANAG Standardization Agreement
TSD Technical Security Division
TRADOC Training and Doctrine Command
TTP actics, techniques and procedures
USNORTHCOM-United States Northern Command
USSS United States Secret Service
UXO unexploded explosive ordnance
VIP Very Important Persons
WIT Weapons Intelligence Teams
WMD weapons of mass destructio

INDEX

A

B

C

D

E

J

L

M

N

O

P

Q

R

S

T

U

V

W

X

Y